THE INCREASING IMPACTS OF HEAT IN THAILAND AND SOUTHEAST ASIA
CHALLENGES AND EMERGING SOLUTIONS

JUNE 2026

ASIAN DEVELOPMENT BANK

6 ADB Avenue, Mandaluyong City, 1550 Metro Manila, Philippines
Tel +63 2 8632 4444; Fax +63 2 8636 2444
www.adb.org

ISBN 978-92-9277-846-0 (print); 978-92-9277-847-7 (PDF); 978-92-9277-848-4 (ebook)
Publication Stock No. SPR260268-2
DOI: http://dx.doi.org/10.22617/SPR260268-2

Notes:
In this publication, "$" refers to United States dollars, "€" refers to euros, "£" to United Kingdom pounds sterling, and "S$" to Singapore dollars.
ADB recognizes "Kyrgyzstan" as the Kyrgyz Republic.

Cover design by Mike Cortes.

Reviewer Perspectives

This primer is timely and policy-relevant. It usefully frames extreme heat as a crosscutting development risk for Thailand and Southeast Asia, couples sectoral impacts with institutional and financing options, and closes with a workable (national) Heat Action Plan framework and city pilots. The narrative is accessible, the graphics are engaging, and the recommendations speak to decision-makers beyond the environment community, public health, labor, urban planning, and finance.

—Andreas Flouris, Professor, University of Thessaly

This report provides a timely, compelling picture of how extreme heat is affecting lives and livelihoods in Southeast Asia. As the authors show, heat stress is deadly and becoming deadlier, particularly for already vulnerable communities. To tackle the systemic danger of heat stress, the authors rightly focus on the need for greater coordination across sectors and institutions, as well as innovative financial mechanisms and public–private partnerships. We at the Rockefeller Foundation welcome the report's practical policy pathways and case-based insights. As countries around the world redesign public health and other systems to help people thrive amid the climate crisis, this report is a valuable contribution to our understanding of heat stress in Southeast Asia and how to confront it.

—Deepali Khanna, Senior Vice President and Head of Asia, The Rockefeller Foundation

This report is an insightful and rigorous contribution, illuminating both the risks of heat stress in Thailand and the opportunities for transformative heat resilience-building. It reminds us that cooling is not a luxury but essential infrastructure, as vital as water, energy, and sanitation. By framing heat resilience-building as a multilevel endeavor that starts with governance and institutional alignment, the report offers a credible and compelling road map toward a cooler, safer, and more equitable future.

—Eleni Myrivili, Global Chief Heat Officer, United Nations Environment Programme (UNDP)

Heat stress is no longer just a climate or health problem, but a governance stress test. But it is also a rare opportunity to unite the whole-of-government, whole-of-society around shared data, resources, mandates, and shared responsibility for protecting human lives.

—Aafreen Siddiqui, UNDP Asia Pacific/ Our Common Air

Extreme heat is emerging as one of Southeast Asia's most silent yet severe climate threats—intensifying health risks, straining water and energy systems, and deepening urban inequalities. While adapting our built environments is necessary, the real solution lies in reimagining them through the lens of nature. Restoring ecosystems, expanding urban forests, and revitalizing wetlands and natural waterways are no longer optional; they are essential climate-resilient infrastructure to address the twin crises of heat and air pollution. This primer offers a compelling synthesis of science, policy, and local action, positioning nature-based solutions as powerful tools for resilience in a rapidly warming world.

—Dindo Campilan, Regional Director for Asia and Hub Director for Oceania,
International Union for Conservation of Nature (IUCN)

This report is very timely on a topic where often there is less attention. However, heat stress is very real and is impacting countries across the world and Thailand is no exception to this. The biggest strength (of) this report is that it tries to take a holistic and multisector approach to this topic as it goes much beyond just tackling climate change. The road map provided is action oriented for various stakeholders. It is an excellent addition to the literature on this topic and will certainly help policymakers and implements in Thailand and Southeast Asia. It has potential for providing lessons and guidance to other developing countries beyond this region that are also dealing with this issue.

—Nishant Jain, Senior Health Specialist, Human and Social Development Sector Office, ADB

Technology may define our era, but I believe overcoming the growing challenge of heat stress—and its compounding effects on air pollution—requires more than innovation. When we connect technology with people, policies, and purpose, we unlock our collective power to drive meaningful heat management and air quality solutions. By working together across our organizations and with strong support from international partners, I'm confident we can build a safer and more resilient future for everyone.

—Siwaporn Rungsiyanon, Specialized Expert on Air Quality and Noise Management, Pollution Control Department Thailand, and Deputy Director General, Asia Center for Air Pollution Research

Heat stress is becoming a growing concern for Thailand, especially Bangkok, but it also gives us a chance to rethink how we care for our city and for one another. By designing cooler streets, supporting vulnerable groups, and creating greener neighborhoods, we can build cities that welcome every resident and every newcomer. A truly livable city is one we build together—for today and for the generations to come.

—Pornphrom Vikitsreth, Advisor to the Governor of Bangkok

CONTENTS

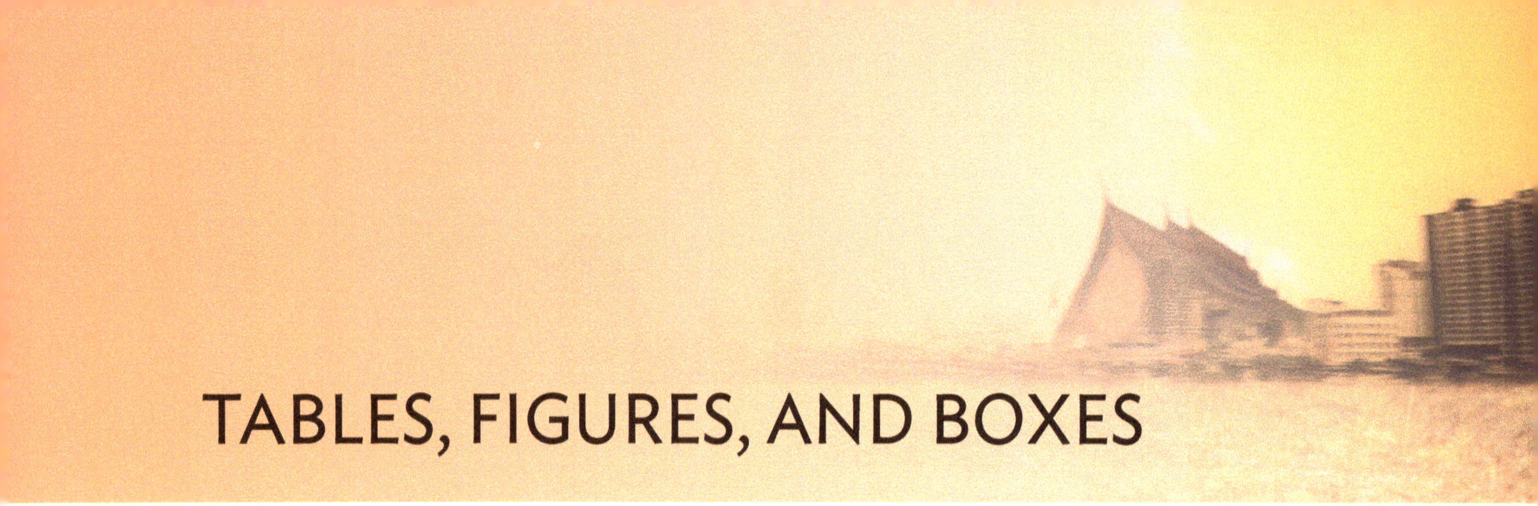

TABLES, FIGURES, AND BOXES

Tables

Figures

Boxes

ACKNOWLEDGMENTS

This special report was prepared by a team led by Anouj Mehta, former Country Director, Thailand Resident Mission and currently Advisor, Office of the Director General, Strategy, Policy, and Partnerships Department, Asian Development Bank (ADB); with ADB consultants Damla Kiremitci, Kewal Thapar, and Sandy Ewart. The report highlights the increasing risk and impacts of heat stress and disseminates information on key regulatory and technical solutions and their practical application by institutions, cities, and governments to reduce its adverse effects. It does not replace or revoke the need to follow any applicable laws or legal provisions and is intended to be used as a guide by ADB's borrowers, clients, and project teams in developing practical responses and pathways to heat stress.

We are grateful for inputs from Amitabh Mehta, founder of Indus Blue Consulting; and peer reviewers, Zonibel Woods, Senior Social Development Specialist (Gender and Development), Gender Equality Division, ADB; Nishant Jain, Senior Health Specialist, Human and Social Development Sector Office, ADB; Eleni Myrivili, Global Chief Heat Officer, United Nations Environment Programme; Andreas Flouris, Professor, University of Thessaly; Dindo Campilan, Regional Director for Asia and Hub Director for Oceania, International Union for Conservation of Nature; A. Siddiqui, Regional Engagement Lead for Digital, Public Sector Innovation and Partnerships, Regional Innovation and Digital Team, United Nations Development Programme Bangkok Regional Hub; and Head of Secretariat, Our Common Air; Siwaporn Rungsiyanon, Director of the Transboundary Air Pollution Division/ Subdivision, Air Quality and Noise Management Division, Pollution Department, Government of Thailand; and Pornphrom Vikitsreth, Advisor to the Governor of Bangkok.

Finally, we would like to extend our deepest thanks to ADB staff and consultants from the Southeast Asia Department, Thailand Resident Mission, and the Department of Communications and Knowledge Management who assisted with the preparation, processing, and production of this publication.

This report was completed with resources and technical support from the technical assistance project, Strengthening the Bio-Circular-Green Economy of Thailand (TA 10079-THA), co-financed by ADB, the Government of Australia, Republic of Korea e-Asia and Knowledge Partnership Fund, and OPEC Fund for International Development.

ABBREVIATIONS

°C	degrees Celsius
ADB	Asian Development Bank
CO_2	carbon dioxide
GDP	gross domestic product
ILO	International Labour Organization
PPP	public–private partnership
UN	United Nations
UNEP	United Nations Environment Programme
UNICEF	United Nations Children's Fund
WEF	World Economic Forum

EXECUTIVE SUMMARY

Heat waves are intensifying. Heat stress is one of the deadliest yet least understood consequences of increasing climate volatility. Heat is no longer a seasonal discomfort that passes by. It has become an intense, relentless, and all-pervading crisis that threatens lives, livelihoods, ecosystems, and our future existence. From scorching urban centers to parched rural landscapes, and from melting glaciers to warmer oceans, heat waves have impacted all geographies and sectors. Heat waves and extreme heat events have intensified in frequency and severity, adversely impacting infrastructure, widening social inequities, and driving up mortality rates. The time to act on this issue has come.

Temperatures are rising fast. The year 2024 was the hottest on record in human history, with temperatures ranging from 1.45°C to 1.60°C, higher than the average in 1850–1900. By 2028, annual global temperatures are projected to rise further to 1.90°C above pre-industrial levels, putting over a quarter of the world's population at risk from heat waves. The effects of a Oceans are already evident: Oceans are getting warmer, ice sheets are shrinking, glaciers are retreating, snow cover is decreasing, sea levels are rising, extreme events like droughts and heat waves are becoming more frequent, and ocean acidification is increasing. All these changes have a profound impact on human well-being, ecosystems, and economies.

Heat stress impacts lives. The most severe impact of extreme heat is on human health, especially the most vulnerable groups, including children and older people, as well as outdoor workers who are most exposed to its debilitating effects. Exposure to high temperatures leads to heat exhaustion and heatstroke and exacerbates chronic conditions like cardiovascular and respiratory diseases. Higher mortality rates due to heat-induced causes, workplace heat injuries, and a rise in infectious diseases are straining public health care services in low-income countries lacking access to reliable electricity and basic water facilities. Extreme heat disproportionately affects the most vulnerable, low-income, and marginalized communities who lack adequate housing, cooling, health care, and emergency support. It has led to widening learning gaps, lower human capital development and productivity, migration, and increased food insecurity.

Extreme heat has multifaceted adverse consequences. Extreme heat is disrupting food systems, straining water supplies, increasing economic losses, reducing labor productivity, and compounding health risks. The global economic and productivity losses due to extreme heat are now at levels that insurance companies no longer consider it as an "invisible peril" and are forced to evaluate the associated investment risks and cascading costs. Extreme heat is now a direct threat to economic competitiveness, depressing outdoor labor output and increasing business costs. For countries reliant on informal workforces, productivity losses could become structurally embedded. By 2025, heat stress was projected to cause $2.4 trillion in productivity losses and $448 billion in annual fixed asset losses for publicly listed companies.

Excessive heat has led to infrastructure vulnerability and compromised the efficiency, integrity, and safety of vital utility services, resulting in frequent power outages, disrupted transportation networks, reduced water availability, and increased maintenance and replacement costs. A warmer planet has caused devastating losses across the entire biodiversity spectrum, from forests to terrestrial and marine organisms and ecosystems—the combined effect of which has triggered multidimensional challenges leading to health deterioration, economic strain, and ecological decline.

Thailand is highly vulnerable. As one of Southeast Asia's most vulnerable countries, Thailand stands on the front lines of climate volatility, with projections indicating a potential loss in gross domestic product of up to 45% by 2048. Agriculture, fisheries, and tourism—the core pillars of the Thai economy and livelihoods—are threatened, as extreme heat, coastal erosion, and flooding intensify. Bangkok and the coastal regions, which are major tourist destinations, are all at high risk from heat-related health, infrastructure, and economic impacts. The Gulf of Thailand, which shares a marine ecosystem with neighboring countries, is under severe ecological stress because of overfishing, marine heat waves, coral bleaching, and pollution. There is an urgent need to reduce the degradation of this shallow, semi-enclosed body of water that provides food security and livelihoods to coastal communities and rest and recreation for tourists.

Scaling up action requires an integrated approach. Several barriers exist that prevent an integrated and systemic response to confronting heat stress and building resilience. These include fragmented sector planning; piecemeal and uncoordinated infrastructure development; people and policy blind spots and lack of awareness; insufficient data, analysis, and monitoring systems; and rising and massive financing gaps. Responses therefore need to integrate policy and projects so that impacts are linked and holistically addressed across different sectors but especially between the highly interlinked themes of air pollution, water management, and heat stress. Financing innovations are especially critical, given the rising fiscal burden of governments, to raise the level of financing for building resilience to extreme heat.

Heat is not just a climate or health challenge—it is also a governance challenge. Managing heat stress requires coordinated action and resources across meteorology, health, labor, urban planning, education, and disaster management, supported by shared data and aligned mandates. Without an integrated governance architecture, responses will remain fragmented and insufficient.

Solutions are emerging. Some nations, cities, and communities have implemented integrated adaptation and mitigation measures—ranging from policy reform to nature-based and technological solutions—to improve public health, enhance biodiversity, and build resilient infrastructure to reduce the human, economic, and environmental toll of heat stress. Experiences from Australia, India, the Kyrgyz Republic, Mexico, the Netherlands, and Singapore offer a wide range of examples affirming that transformational solutions are possible and provide models to build upon and scale up.

Creating a road map to a heat action plan. Building on international good practices, this paper outlines the causes and consequences of heat stress resulting from global temperature rise and its impacts on health, economics, and society. It discusses adaptation and mitigation options, highlighting technical innovations, institutional frameworks, and financial approaches that can be adopted to alleviate adverse heat stress impacts. It concludes by proposing a road map for developing a heat action plan for Thailand. The road map prioritizes measures to protect vulnerable populations, mobilize public and private finance, and accelerate coordinated cross-sector solutions. It calls for strong leadership from government agencies, active engagement of international financial institutions, and collaboration with communities and the private sector to ensure an inclusive, evidence-based, and sustainable response to heat stress. With bold policy shifts, inclusive planning, and sustained investment, we can move from climate crisis to resilience—ensuring that communities adapt and withstand the next wave of heat. The time to act and scale is now.

RISING GLOBAL TEMPERATURES AND HEAT STRESS

Trends in Temperature Rise

As greenhouse gas emissions blanket the Earth, they trap the sun's heat, preventing it from escaping back into space. This buildup of heat in the atmosphere leads to global warming, which disrupts weather patterns and causes long-term changes to the climate. As a result, the planet is now warming at a rate faster than at any other time in recorded history, intensifying the urgency to address this global crisis.

In 2011–2020, the Earth's surface was 1.1 degrees Celsius (°C) warmer than pre-industrial levels (1850–1900), and atmospheric carbon dioxide (CO_2), methane, and nitrous oxide reached their highest levels in millions of years.[1] Figure 1 provides a visual representation of these global warming trends.

Figure 1: Abnormal Global Warming, 1880–2024

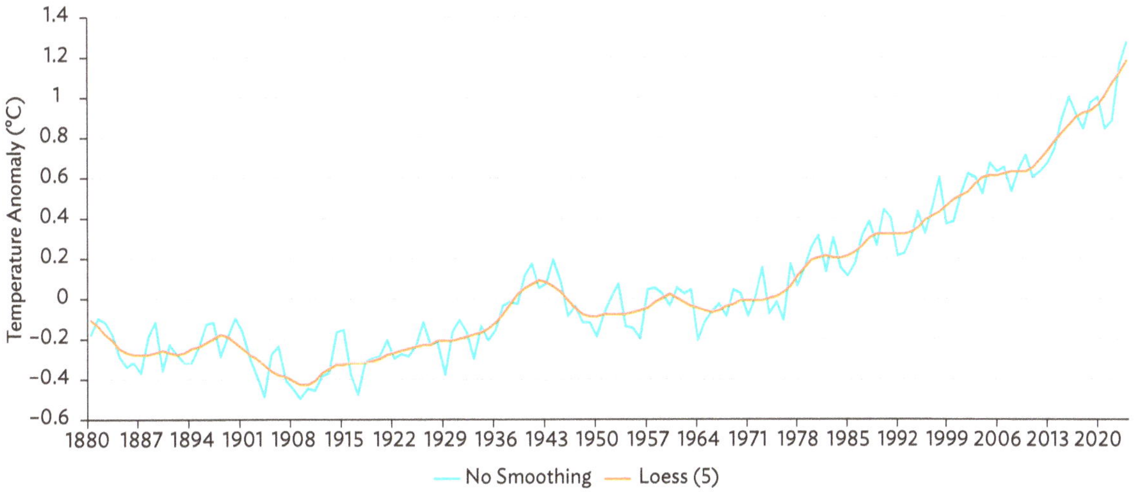

Source: National Aeronautics and Space Administration (NASA). Global Temperature—Earth Indicator. Global Land-Ocean Temperature Index (accessed 10 October 2025).

1 Intergovernmental Panel on Climate Change (IPCC). 2023. Summary for Policymakers. In Core Writing Team, H. Lee, and J. Romero, eds. *Climate Change 2023: Synthesis Report. Contribution of Working Groups I, II and III to the Sixth Assessment Report of the Intergovernmental Panel on Climate Change*. pp. 1–34.

A recent report by the World Meteorological Organization has shown 2024 as the warmest year on record globally.[2] The report highlights that global warming is accelerating exponentially. In the last 10 years, each successive year has been the hottest on record. Ocean temperatures have hit new highs for 8 consecutive years; Arctic and Antarctic sea ice is shrinking to record lows; and glaciers are rapidly retreating, with the largest reported loss of ice in the last 3 years. Consequently, sea levels are rising twice as fast as when satellite tracking began, threatening coastal communities worldwide.

Setting new records, the global mean near-surface temperature for each year in 2024–2028 is predicted to be between 1.1°C and 1.9°C, higher than the 1850–1900 average (footnote 2). At 2°C global warming, the population exposed to at least 1 day per year exceeding the lethal threshold of compound heat and humidity could reach 2.3 billion people.[3]

Changes in the climate have caused widespread adverse impacts, with economic damage evident in agriculture, forestry, energy, tourism, and urban infrastructure. Vulnerable populations, particularly in low-income regions, bear a disproportionate share of these impacts, further exacerbating inequalities. Approximately 3.3 billion–3.6 billion people live in highly vulnerable contexts and are experiencing compounded impacts (footnote 1).

Heat Stress Impacts

Heat stress is often used as an all-encompassing phrase for what are essentially similar but also different forms of response to extreme heat.[4] Table 1 provides clearer definitions of heat stress, heat strain, and heat illness.

Table 1: Heat Stress, Heat Strain, and Heat Illness Definitions

Term	Definition	Key Elements
Heat stress	The total environmental and metabolic heat load placed on the body	• Environmental heat (temperature, humidity, radiant heat, air movement) • Metabolic heat (body heat generated during physical activity)
Heat strain	The physiological response of the body to heat stress, including cardiovascular, thermoregulatory, and hydration-related changes	• Increased heart rate • Sweating • Rising core and skin temperature • Fluid and electrolyte shifts
Heat illness	A spectrum of medical conditions resulting when heat strain exceeds the body's capacity to compensate	• Heat rash • Heat cramps • Heat syncope (fainting) • Heat exhaustion • Heatstroke (medical emergency)

Sources: Authors' analytics and synthesis of definitions from American Conference of Governmental Industrial Hygienists (ACGIH). 2022. Health Stress and Strain; Centers for Disease Control and Prevention (CDC) and National Institute for Occupational Safety and Health (NIOSH). 2016. *Criteria for a Recommended Standard: Occupational Exposure to Heat and Hot Environments*; CDC and NIOSH. 2024. Heat Stress and Workers Safety; Occupational Safety and Health Administration (OSHA). 2023. *OSHA Technical Manual: Heat Stress*; World Health Organization (WHO). 2024. Heat and Health. Fact Sheet. 28 May.

---

2 World Meteorological Organization. 2025. *State of the Global Climate 2024*.

3 D. Li, J. Yuan, and R. E. Kopp. 2020. Escalating Global Exposure to Compound Heat-Humidity Extremes with Warming. *Environmental Research Letters*. 15 (6). 064003.

4 In this document, the terms *heat stress* and *extreme heat* are used interchangeably. While heat stress refers to a physiological response and extreme heat to a broader climatic phenomenon, this approach is intended to support a wider audience's general understanding of the key issues and maintain consistency with the sources used in developing the document.

Rising Heat Waves

Annual deaths due to extreme heat totaled 489,000 between 2000 and 2019—more than floods, hurricanes, earthquakes, and wildfires combined[5]—with 45% of these deaths in Asia and 36% in Europe.

As global temperatures continue to rise, heat waves have become more frequent and intense; global surface temperature will continue to increase until at least mid-century under all emissions scenarios. Global warming of 1.5°C and 2.0°C will be exceeded during the 21st century unless deep reductions in CO_2 and other greenhouse gas emissions occur in the coming decades.[6]

This is posing serious risks to public health, economies, and social stability. The World Economic Forum (WEF) notes that, per United Nations Environment Programme (UNEP) data, rising temperatures will lead to a surge in deaths especially in vulnerable population, with a 370% projected increase in annual heat-related deaths in older populations if global temperatures rise by 2.0°C.[7]

Overlaying climate-related hazards and multidimensional poverty data, the 2025 Multidimensional Poverty Index Report has established high heat as the most widespread hazard confronting poor people globally. Specifically, of the 887 million poor people living in regions experiencing at least one of four climate hazards, 608 million are exposed to high heat, 577 million to air pollution, 465 million to floods, and 207 million to drought.[8]

Health Impacts

Extreme heat can lead to health issues, ranging from mild discomfort to life-threatening heatstroke and organ failure. Heat waves also exacerbate the spread of vector-borne diseases, such as malaria and dengue, as well as worsen noncommunicable diseases like heart conditions. Extreme heat significantly increases mortality among young children and older people, primarily because of cardiovascular stress, and causes a rise in acute enteric infections.

Prolonged extreme heat also undermines human nutrition and food security by reducing the nutritional value of cereal grains and decreasing livestock productivity, including the production of eggs, milk, and meat. Furthermore, it adversely affects human productivity and increases the occupational health risks particularly of those engaged in outdoor work such as market vendors, farmers, fishers, and construction workers who bear the greatest burden because of limited protection and resources.

Figure 2 illustrates the wide-ranging impact of heat stress on human health and well-being.

Uneven Geographic Impacts

Extreme heat impacts, which are determined by a combination of environmental, social, economic, and physiological factors, vary in extent across geographies. For instance, while the International Labour Organization (ILO) projects a worldwide loss of 2.2% in working hours by 2030, this loss will be much higher in South Asia and Western Africa (around 5.0%) compared to European subregions (0.1%).[9] The concentration of labor-intensive

5 T. Crowfoot. 2025. Extreme Heat: What to Know About the Deadliest Climate Risk of Our Time. *WEF*. 30 July.

6 IPCC. Summary for Policymakers. In V. Masson-Delmotte et al., eds. *Climate Change 2021: The Physical Science Basis. Contribution of Working Group I to the Sixth Assessment Report of the Intergovernmental Panel on Climate Change.*

7 WEF. 2024. *Quantifying the Impact of Climate Change on Human Health: Insight Report*. In collaboration with Oliver Wyman.

8 Oxford Poverty and Human Development Initiative and United Nations Development Programme. 2025. *Global Multidimensional Poverty Index 2025—Overlapping Hardships: Poverty and Climate Hazards*.

9 ILO. 2024. *Ensuring Safety and Health at Work in a Changing Climate: Global Report*.

Figure 2: The Wide Impact of Heat Stress

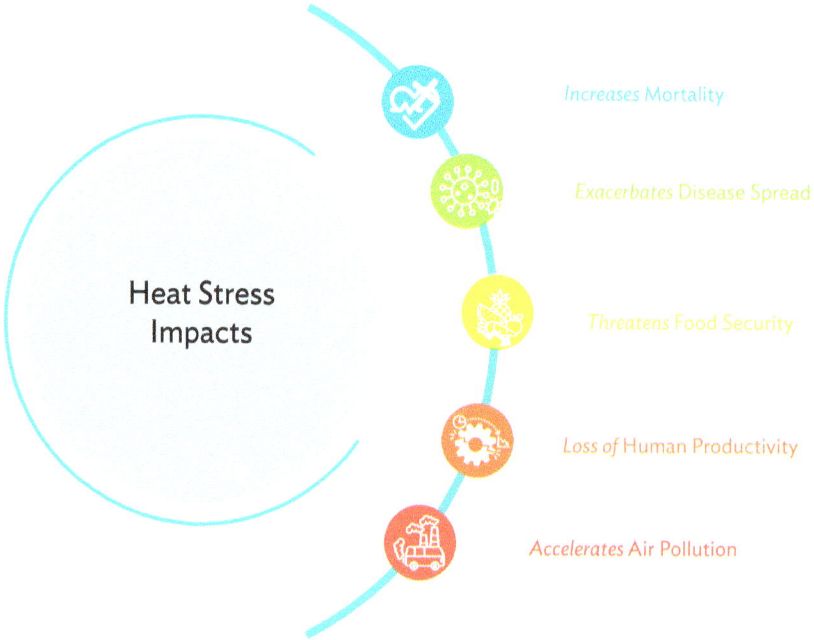

Source: Authors.

sectors, such as agriculture, in tropical and subtropical regions makes them particularly vulnerable. Poorer regions face disproportionate impacts, while urban areas in developing countries experience four times higher economic losses than their northern counterparts because of inadequate infrastructure and climate adaptation measures.

Vulnerable Group Impacts

Across all geographies, vulnerable groups are most impacted. These include people in poverty, older adults, individuals with preexisting health conditions, women, pregnant women, infants, children, and outdoor workers. Citing various reports, the United Nations (UN) Secretary-General has highlighted worsening trends and projections on these groups' exposure and vulnerability to extreme heat impacts.[10] From 24% in 2020, nearly all the world's 2.2 billion children under 18 could be exposed to high heat waves by 2050.[11] Heat-related mortality for individuals over 65 years increased by 85% between 2000 and 2004 and between 2018 and 2022 (footnote 10).

A 1.0°C temperature rise increases poverty by 9.1% and the Gini inequality index by 0.8%. In an average year, heat stress causes poor rural households to lose 5.0% of their annual income. Many parts of Asia and North Africa experience school closures due to extreme heat. But as it is hard to learn in extreme heat, even if schools are not closed, extreme heat disrupts education, leading to widening learning gaps and lower human capital development. In 2024, more than 80 million students were affected by school closures (footnote 10).

[10] United Nations. 2024. *United Nations Secretary-General's Call to Action on Extreme Heat*.

[11] United Nations Children's Fund (UNICEF). 2022. *The Coldest Year of the Rest of Their Lives. Protecting Children from the Escalating Impacts of Heat Waves*.

As of 2024, more than 70% of the global workforce (2.4 billion) was at high risk from extreme heat, with excessive heat causing 22.85 million occupational injuries and 18,970 deaths annually.[12] Furthermore, 9 out of 10 worker exposures to excessive heat and 8 out of 10 occupational injures linked to excessive heat *occur outside of a heat wave,* underscoring workers' lack of heat protection. Workers in Africa (92.5%), the Arab states (83.6%), and Asia and the Pacific (74.7%) are most exposed to excessive heat. The figures are above the global average of 71.0% (footnote 9). Workplace heat injuries with related losses could amount to as much as 1.5% of national gross domestic product (GDP) in some countries.

Wet-Bulb Temperatures

These temperatures, providing combined readings of heat and humidity, have reached unbearable levels in parts of South Asia, the Middle East, and the African Sahel region and are predicted to become more intense. If the wet-bulb temperatures exceed 35°C for over 5 hours, the human body loses its capacity to cool itself through sweating. This could result in overheating, leading to organ failure, brain damage, and even death in the absence of timely medical intervention. According to the UN University–Institute for Environment and Human Security, wet-bulb temperatures have crossed this limit at two weather stations—one in the Persian Gulf and the other in the Indus River Basin.[13] Research shows that, by 2070, these threshold limits will regularly be exceeded in South Asia and the Middle East, and by 2100, Southeast Asia will face the highest risk of heat-related mortality under 1.5°C to 3.0°C warming scenarios (footnote 10).

Figure 3 shows when these limits were breached in parts of Asia (left graph) and the cities where these breaches are predicted to be highest under different warming scenarios (right graph).

Figure 3: Temperatures in Asia Exceeding the Wet-Bulb Temperature of 35°C

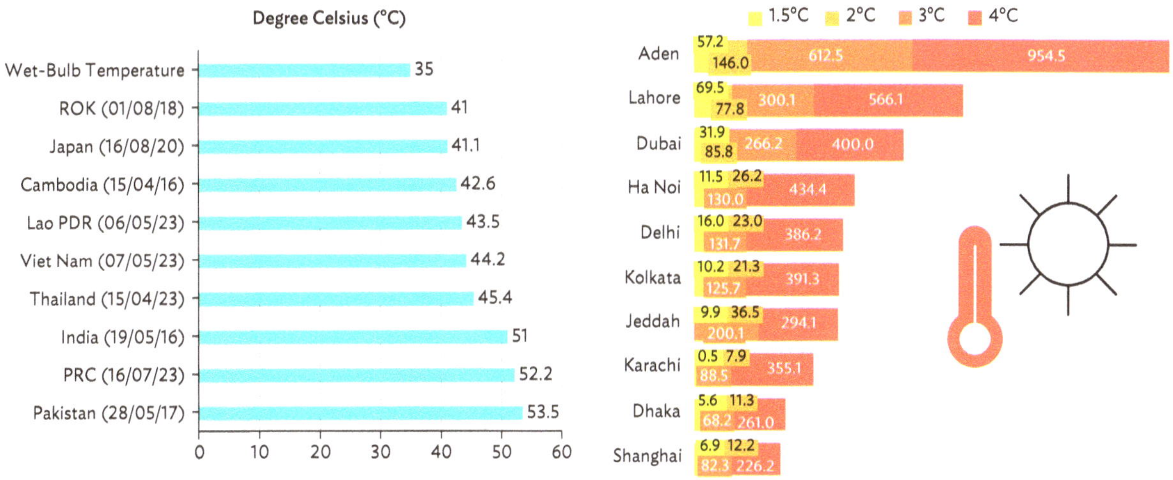

PRC = People's Republic of China, ROK = Republic of Korea, Lao PDR = Lao People's Democratic Republic.

Sources: (i) Left graph: Authors, using data from Extreme Temperatures Around the World (accessed 26 June 2025); and (ii) Right graph: Adapted from K. Buchholz. 2023. Where Will Wet-Bulbs Turn Unbearable? *Statista*. 11 October.

12 A. Flouris et al. 2024. *Heat at Work: Implications for Safety and Health—A Global Review of the Science, Policy and Practice*. ILO.

13 EurekAlert!. 2023. UN University Report Warns About Risk Tipping Points with Irreversible Impacts on People and Planet. *News release.* 25 October.

Economic Impacts

The economic impacts of climate-induced heat stress are also significant. In 2024, weather-related extreme events caused $241.95 billion in global economic losses,[14] while heat-related mortality and associated economic losses for 2019–2023 cost $199 billion annually, up by 179% since 2000.[15] The ILO projects that, by 2030, global labor productivity losses due to high temperatures could reach 2.2%[16]—equivalent to 80 million full-time jobs (footnote 9). This will reduce global GDP by $2.4 trillion, nearly 10 times the 2024 reported economic loss (footnote 10).

Record-Breaking Heat in Southeast Asia

In April 2024, Southeast Asia experienced unprecedented heat waves. Temperatures soared to 45.4°C in Tak, Thailand, and 44.1°C in Thanh Hoa, Viet Nam. Since the 1960s, Southeast Asia's annual mean temperature has risen by 0.14°C–0.20°C per decade, with heat waves becoming more frequent and severe.[17] Climate data show that average temperatures in Southeast Asia have steadily increased, with projections indicating further warming. Projections warn that, without emission reductions, temperatures in major Asian cities could rise by 4.0°C, surpassing critical thresholds for human survival.[18] Urban areas such as Bangkok, Jakarta, and Manila are particularly at risk because of the urban heat island effect.

Rising CO_2 levels lower crop nutrition, escalating malnutrition—the 2015–2016 El Niño left 20.5 million people in Southeast Asia food-insecure, and 5.9 million children underweight, foreshadowing future crises (footnote 15). By 2050, studies project a 700% global increase in the number of urban poor living in extreme heat conditions, driven by rapid urbanization, insufficient infrastructure, and limited adaptive capacity, with the largest increases across West Africa and Southeast Asia.[19]

Despite the growing threat, many Southeast Asian cities lack adequate heat risk assessments and early warning systems. Urban infrastructure fails to provide sufficient cooling, and regulations protecting outdoor workers from extreme heat are either weak or unenforced. A proactive approach to policy reform and infrastructure investment is critical.

14 Centre for Research on the Epidemiology of Disasters. 2025. *2024 Disasters in Numbers*.

15 Lancet Countdown. Health and Heat: Heat-Related Mortality (accessed 11 October 2025).

16 When temperatures rise above 24°C–26°C, labor productivity begins to decline. At 33°C–34°C, productivity drops by 50%.

17 S. C. Kwan and D. McCoy. 2022. Climate Change in South East Asia: Where Are We and What Are We Bound For? *United Nations University*. 28 June.

18 Asian Development Bank (ADB). Rising Above the Heat: Strengthening Women's Resilience to Heat Stress.

19 Urban Climate Change Research Network. 2018. *The Future We Don't Want: How Climate Change Could Impact the World's Greatest Cities*.

THAILAND: VULNERABILITY TO EXTREME HEAT

Heat Trends and Projections

Thailand's average maximum temperatures have risen by ~1.0°C since the early 1980s (Figure 4), with more frequent and intense heat waves. The country ranked among the top five countries expected to have the largest decline in GDP because of climate volatility by 2048.[20]

Thailand faces escalating challenges from changes in the climate, with temperatures projected to rise by up to 3.2°C by 2090. Its economy relies on agriculture, industry, services, and tourism, supporting an estimated population of 71.62 million (2025). Yet, agriculture, water management, health, and tourism are already vulnerable, with economic losses from farmland value and output alone expected to exceed $94 billion by 2050 under a high-emissions scenario.[21] These projections demand urgent and practical adaptation measures to mitigate impacts and safeguard livelihoods.

Figure 4: Annual Mean Maximum Temperature in Thailand, 1982–2020

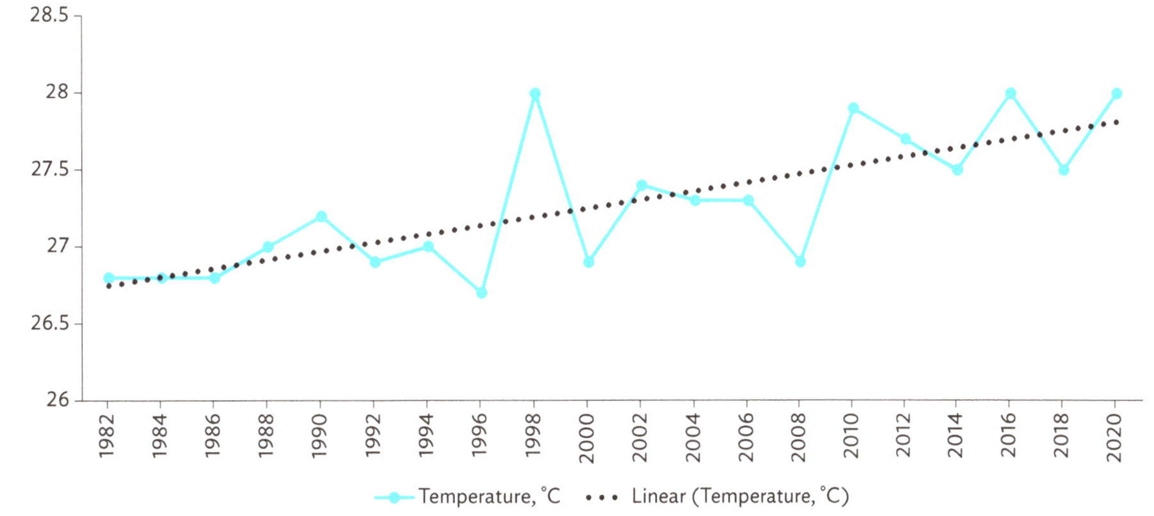

Source: Thai Meteorological Department. 2022. Annual Mean Maximum Temperature in Thailand (accessed 17 June 2025).

[20] World Bank. 2022. Thailand Rural Income Diagnostic: Challenges and Opportunities for Rural Farmers.

[21] World Bank Group and ADB. 2021. Climate Risk Country Profile: Thailand.

The 2023 El Niño event significantly intensified drought conditions across Thailand, disrupting water availability, agriculture, and energy supply. As El Niño weakened the southwest monsoon system, the country experienced reduced rainfall during the typically wet months, leading to abnormally dry conditions, particularly in the central and northeastern regions. Major reservoirs such as Bhumibol and Sirikit saw critically low water levels, prompting water use restrictions and affecting irrigation-dependent crops like rice and sugarcane. The Royal Irrigation Department and the Ministry of Agriculture issued drought warnings earlier than usual, underscoring the severity and geographic extent of the crisis.

Simultaneously, El Niño contributed to record-breaking heat across Thailand, with several provinces experiencing prolonged periods of extreme temperatures exceeding 40°C. In April 2023, parts of the country recorded their highest temperatures in decades, intensifying evapotranspiration and exacerbating already dry soil conditions. The combined effect of reduced rainfall and extreme heat placed significant stress on public health, particularly among vulnerable populations, and led to increased electricity demand due to the widespread use of air conditioning. These compound climate impacts underscored the need for more robust drought preparedness and adaptive water management strategies in the face of recurring El Niño episodes.

Marine Heat Waves

Marine heat waves, particularly on the Gulf of Thailand, increase the vulnerability of key sectors—agriculture and tourism—or impact due to changes in the climate, including sea level rise, flooding, and extreme weather events.[22] Bangkok, with a population of over 17.4 million and a tourist hub receiving over 30 million visitors in 2024, faces a dangerous future as temperatures rise. With just a 1°C increase in temperatures, annual deaths due to heatstroke doubles, worker output declines by over 3%, and households will spend 30% more on energy, thus contributing more CO_2 into the atmosphere and deepening the climate crisis.

The Gulf of Thailand, a vital marine ecosystem for Cambodia, Malaysia, Thailand, and Viet Nam, faces severe ecological and environmental pressures that impact local livelihoods and the regional economy. Its shallow, semi-enclosed feature with minimal circulation makes it susceptible to rapid warming, pollution, and habitat degradation. Record-breaking marine heat waves have raised surface temperatures to over 32°C, leading to extensive coral bleaching and fish mortalities. Catch per unit effort has plunged by 90% since the 1960s because of overfishing and illegal, unreported, and unregulated practices, thereby undermining food security and incomes for coastal communities. The degradation of the marine environment threatens Thailand's tourism sector that is dependent on healthy coastal and marine biodiversity. The competing uses of resources, coupled with inadequate waste management and industrial pollution, further exacerbate the marine ecology stress.

Urban Heat

Concrete-dominated cities, like Bangkok and Chiang Mai, experience significantly higher temperatures than surrounding rural areas, with local temperatures rising by 2°C–4°C because of the urban heat island effect. The abundance of heat-absorbing surfaces such as asphalt and concrete, combined with dense infrastructure, traps heat during the day and releases it slowly at night, preventing nighttime cooling and intensifying overall heat exposure for city residents.

In informal settlements, the lack of green spaces and poor ventilation further increase vulnerability to extreme heat. These areas often lack adequate urban planning, with tightly packed housing that limits airflow and reduces the cooling benefits of vegetation. As a result, residents face greater health risks during heat waves, particularly in the absence of reliable access to cooling, health care, or other adaptive resources.

22 World Bank Group. Climate Change Knowledge Portal. Thailand (accessed 25 October 2025).

Socioeconomic and Sector Impacts

Although *heat stress* is often discussed as a global concept, its interpretation varies among individuals and organizations and to what the term is being applied. Table 2 presents a brief list of commonly used terms and definitions, summarized for clarity to ensure a consistent understanding of the terminology used throughout this report.

Table 2: Heat Metrics—A Quick Guide

Heat Metric	What It Measures and Definition	What It Is Used For	Occupational Use?
Dry-bulb temperature	Standard air temperature (what a regular thermometer reads)	Basic air temperature measurement	Useful, but insufficient for heat-stress decisions
Wet-bulb temperature (or natural wet-bulb)	Temperature accounting for evaporative cooling (influenced by humidity and wind)	Assess cooling ability of the environment; track strain risk when sweat cannot evaporate well	Important input; part of wet bulb global temperature; not a stand-alone limit in workplaces
Wet-bulb global temperature	Composite index including wet-bulb, dry-bulb, and radiant heat (via black globe)	Gold-standard index for occupational heat stress, training, and military use	Primary recommended metric;[a] used for work and/or rest guidance
Heat index	Combines air temperature + humidity to reflect "feels-like" heat in shade with light wind	Public weather warnings and community heat-safety messaging	Not recommended for most occupational decisions (does not include radiant heat or workload)
Universal thermal climate index	Advanced heat-stress index using physiology-based models (air temperature, humidity, radiation, wind)	Public health, urban planning, environmental heat exposure modeling	Emerging tool; useful supplement, but wet bulb global temperature still prefers workplace policy

[a] By reputable technical and research agencies, including the United States government's National Institute for Occupational Safety and Health and Occupational Safety and Health Administration and the American Conference of Governmental Industrial Hygienists.
Source: Authors' analytics.

Health

The most severe impact of extreme heat is on human health, as exposure to high temperatures leads to heat exhaustion and, heatstroke, and exacerbates chronic conditions like cardiovascular and respiratory diseases.[23] Using the "universal thermal climate index" expounded in Table 3, Figure 5 illustrates the impact of extreme heat on the human body. The universal thermal climate index is a metric that quantifies how people feel in urban outdoor spaces, considering the combined effects of temperature, wind, humidity, and sunlight and/or radiation.

Episodes of extreme heat are increasingly occurring alongside high concentrations of air pollutants, particularly in urban areas where both stressors are becoming more frequent and intense. This dangerous combination leads to compound exposures that significantly elevate the risk of cardiorespiratory illness and death, far exceeding the health impacts posed by heat or pollution alone.

[23] World Health Organization (WHO). 2024. Heat and Health. *Fact sheet.* 28 May.

Table 3: Universal Thermal Climate Index and Corresponding Physiological Responses

Temperature	Heat Stress Level	Physiological Responses
>46°C	Extreme	Associated with an increased risk of heat-related health issues, and precautions are necessary to avoid heat-related illnesses.
+38°C–46°C	Very strong	Increased strain on the body's thermoregulatory mechanisms. Prolonged exposure could lead to mild thermal strain.
+32°C–38°C	Strong	Temperature increases after 30 minutes' exposure. Heat exhaustion
+26°C–32°C	Moderate	Body's ability to regulate temperature is generally effective. Some discomfort (fatigue, heat cramps). No major physiological strain.
+9°C–26°C	None	Individuals can maintain a thermal equilibrium without strain on the body.

Source: Adapted from *Copernicus Climate Change Service*. 2024. Heat Stress: What Is It and How Is It Measured? *News*. 22 May.

Figure 5: Health Impacts of Extreme Heat

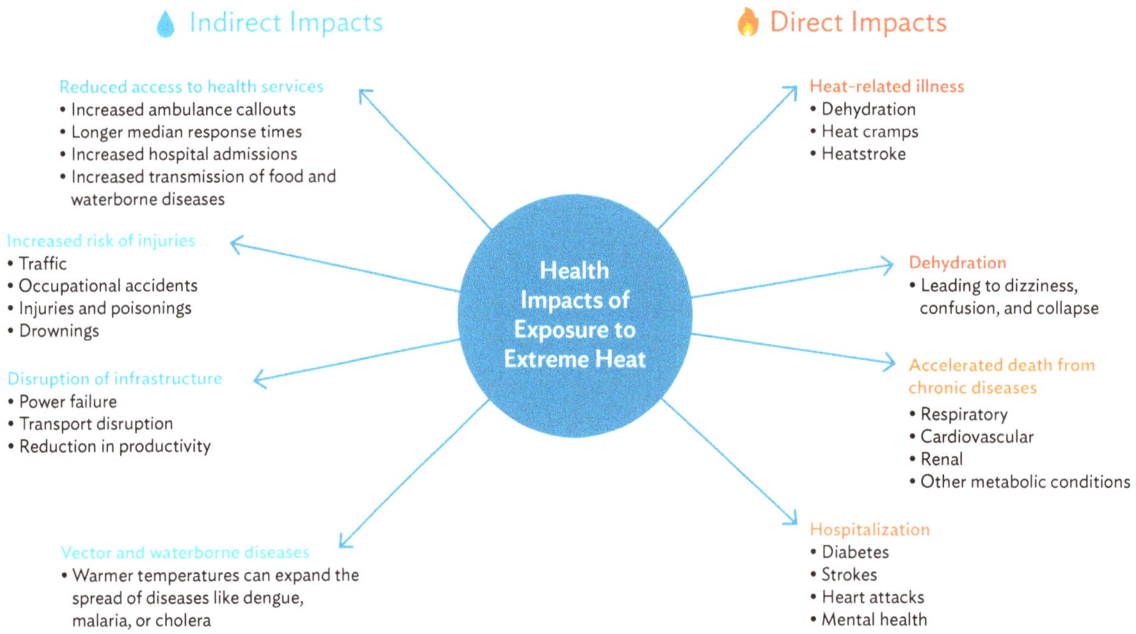

Source: Authors' analysis, based on various research findings.

While not facing the same utility shortages as many low-income countries, climate vulnerabilities persist in Thailand. An assessment of four health care institutions (Rayong Hospital, San Sai Hospital, Suthep Community Health Center, and Bueng Yitho Medical and Rehabilitation Center) found exposure to flooding, water scarcity, and air pollution—underscoring the need for climate-resilient infrastructure.[24]

[24] T. Sudhipongpracha. 2023. *Preliminary Assessment Report: Green Viability and Climate Vulnerability Assessments of Healthcare-Related Facilities in Thailand*. United Nations Population Fund (UNFPA).

Labor and Productivity

Heat stress is also taking a toll on workers. In Bangkok, heat and humidity are a major drag on worker productivity, causing overall losses valued at 5% of output in a typical year, or $8.6 billion. Without action to reduce emissions or adapt to the changing climate, heat- and humidity-related labor productivity losses could reach almost $16 billion by 2050, or 6% of total economic output, and a value equivalent to a quarter of prepandemic tourist spending across all of Thailand.[25] Outdoor workers (e.g., in agriculture and construction) lose up to 20% productivity during peak heat.

Agriculture

Rising temperatures negatively impact crop yields by causing heat stress, reducing the productivity of staple crops like wheat, rice, and maize. Livestock are similarly affected, with reduced fertility, lower milk, and meat production and higher mortality rates under extreme heat conditions.

Thailand's agriculture sector, particularly rice production, is highly vulnerable to intensifying heat stress. Rising temperatures are projected to reduce rice yields, impair farmers' health, degrade soils, and increase postharvest losses along the supply chain. Studies show that prolonged exposure to daytime temperatures of 35°C and nighttime levels of 30°C leads to rice sterility and depressed grain formation, while a 1°C rise in maximum daily temperature can reduce yields by up to 10%. In addition, heat stress undermines grain quality, driving down market prices and reducing farmer incomes. By 2028, cropland suitable for rice in Thailand could decline by 7% compared with 2021 levels, further eroding national production capacity.[26]

The compounding effects of changes in the climate amplify these challenges. Extreme heat accelerates soil degradation through cracking, compaction, and biodiversity loss, while floods, droughts, and sea level rise contribute to topsoil erosion, moisture depletion, and salinity intrusion. Heat waves and droughts also distort crop shapes, making produce less marketable, while high humidity speeds up bacterial growth, leading to spoilage during storage and transport. These stresses collectively threaten food security, rural livelihoods, and Thailand's position as a global rice exporter.

Tourism

With heat waves becoming longer and more intense, the tourism sector—which directly and indirectly contributed about 20% to Thailand's GDP in 2019[27]—is increasingly vulnerable. Thailand's traditional peak season, from November to February, offers relatively cooler weather. However, the hot season (March to May) has become almost unbearably warm in many regions. In April 2023, temperatures in parts of northern Thailand exceeded 44°C.[28] Popular outdoor attractions like temples, beaches, and national parks become uncomfortable or even unsafe during heat waves, leading to a decline in visitor satisfaction and an increase in heat-related health incidents among tourists.

The economic impact is widespread. Hotels and resorts face mounting costs due to higher energy and water demands, while vulnerable ecosystems like coral reefs and forests suffer from heat stress, coral bleaching, and fire risk—reducing the appeal of marine and nature-based tourism. Coral reefs in the Andaman Sea and the Gulf of Thailand have shown signs of stress during marine heat waves, with bleaching events recorded as recently as 2024.[29]

25 Atlantic Council Climate Resilience Center (formerly Adrienne Arsht-Rockefeller Foundation Resilience Center). 2021. City Report—Bangkok, Thailand. In *Hot Cities, Chilled Economies: Impacts of Extreme Heat on Global Cities.*

26 E. Ludher. 2025. The Critical Impact of Extreme Heat on Rice Production in Southeast Asia. *ISEAS Perspective.* No. 48. ISEAS - Yusof Ishak Institute.

27 ADB. 2020. *Report and Recommendation of the President to the Board of Directors: Proposed Countercyclical Support Facility Loans to the Kingdom of Thailand for the COVID-19 Active Response and Expenditure Support Program.*

28 *Bangkok Post.* 2023. Searing Heat, A Wake-up Call. Editorial. 7 May.

29 T. Yeemin. 2024. Member's Report (Thailand). 38th General Meeting of the International Coral Reef Initiative. Jeddah. 9–13 September.

Nature-based tourism in northern and central Thailand is also at risk. Increased heat and reduced rainfall negatively affect the country's forests, wildlife, and waterfalls, which are central to national park tourism. In some areas, forest fires—exacerbated by drought—have led to park closures, disrupting travel plans and damaging Thailand's reputation as a nature destination.[30] Seasonal unpredictability has also made it harder for tour operators to plan ahead, impacting their reliability and profitability.

Social Sector

Extreme heat amplifies social inequalities, disproportionately affecting low-income and marginalized populations who often lack access to adequate cooling systems, quality health care, and reliable emergency support. These communities may live in poorly insulated housing or urban heat islands, where temperatures are significantly higher than in wealthier neighborhoods. Without access to air conditioning or safe cooling centers, they face increased risks of heat-related illnesses and fatalities.

Box 1 highlights that heat risks are exacerbated by the convergence of three compounding factors: rising heat, high air pollution, and rapid aging.

Box 1: Convergence of Three Reinforcing Risks

Thailand faces a unique convergence of three reinforcing risks that together amplify climate, health, and economic impacts:

- **Rising heat.** Average maximum temperatures have already increased and are projected to rise further, with more frequent and intense heat waves, particularly in Bangkok and secondary cities. Nighttime temperatures increasingly remain high, limiting physiological recovery.
- **High air pollution.** Seasonal fine particulate matter peaks (PM2.5) from transport, industry, and agricultural burning mean that many communities experience simultaneous exposure to extreme heat and hazardous air, significantly increasing cardiorespiratory risks.
- **Rapid aging.** By 2030, nearly one in three Thais will be aged 60 or older, a group highly vulnerable to heat stress, chronic disease aggravation, and limited mobility—especially in poorly cooled, low-income, and rural settings.

This "triple exposure" makes it essential for Thailand to treat heat resilience, air quality management, and aging policy as a single, integrated agenda—linking climate, health, social protection, urban planning, and digital early warning systems under a coherent national framework.

Source: A. Siddiqui, United Nations Development Programme Asia Pacific.

As extreme heat events become more frequent, public health risks are amplified—especially for older adults, who are more vulnerable to heat stress, dehydration, and heatstroke. Those in rural or low-income communities are particularly at risk, as they often live in poorly ventilated housing or lack access to air conditioning. By 2030, nearly 30% of the population will be aged 60 or older, placing significant additional pressure on health care and social systems. Hospitalizations for heat-related conditions, along with chronic diseases, worsened by high temperatures such as cardiovascular and respiratory issues, are expected to rise. Caregivers and health workers are compelled to navigate the complexities of geriatric care and climate adaptation, without sufficient training or resources.

30 *The Nation*. 2023. Thailand Closes Eight National Parks in the North to Prevent Forest Fires. 18 February.

Economically, this convergence presents challenges to Thailand's productivity and social support systems. As the working-age population shrinks, there are fewer people available to support both the aging population and the necessary climate adaptation infrastructure. Older adults, who may once have contributed to agricultural or informal labor, are less able to participate during extreme weather conditions. This combination of climate vulnerability and demographic imbalance threatens long-term economic resilience, requiring urgent investment in climate-smart health systems, age-friendly urban planning, and heat-resilient housing.

Rising temperatures in Thailand exacerbate gender inequalities, disproportionately impacting women, particularly those in low-income or rural communities. Many women work in informal or outdoor sectors—such as street vending, agriculture, or tourism—where they are more directly exposed to extreme heat without adequate protection or benefits like paid leave or health care. Prolonged exposure increases the risk of heat-related illnesses, while limited access to air-conditioned environments or cooling infrastructure further intensifies their vulnerability. Within the household, women often carry the burden of caregiving for children, older people, and the sick—roles that become more demanding during heat waves. Women are also frequently responsible for tasks such as cooking and water collection, which become more physically taxing and time-consuming in hotter conditions, especially when clean water supplies are strained by droughts.

Increasing heat is having a growing impact on the health, safety, and learning outcomes of school-aged children, with predictions of nearly 2.2 billion children likely to be exposed to frequent heat waves by 2050.[31] Children are particularly vulnerable to heat-related illnesses like heat exhaustion and dehydration because of their underdeveloped thermoregulation. Many Thai schools, especially in rural or low-income areas, lack adequate cooling infrastructure. Classrooms can reach dangerously high temperatures during the hot season (March to May), leading to decreased concentration, fatigue, and, in some cases, heat-related illness. Research also shows that high temperatures impair cognitive performance and exam results among children, raising concerns about long-term educational disparities.

Beyond classroom discomfort, heat waves contribute to broader socioeconomic stress that affects children's well-being. School closures due to extreme heat can interrupt learning and increase dropout risk, especially for students in vulnerable communities with limited digital access to online tutoring. Heat waves will also lead to increased nutritional challenges and food insecurity arising from crop failures, and will threaten children's access to food and water. Heat has also been linked to mental health problems in children and adolescents (footnote 31).

Without urgent investment in climate-resilient schools and social protection, the educational and developmental rights of children will continue to be threatened by rising temperatures.

Urban Infrastructure and Energy

Heat has a profound impact on infrastructure with devastating effects. The impacts are due to thermal expansion and contraction of materials, increased demand for services, and a reduction in material performance leading to structural damage, disruptions, increased maintenance costs, and reduced operational efficiency. It compromises the safety, efficiency, and longevity of air, road, and rail transportation systems; and strains energy generation, transmission, and distribution systems leading to outages. Due to extremely hot weather conditions, Thailand experienced peak electricity demand of 36,699.9 megawatts on 29 April 2024, at 9:00 p.m., the highest in the history of electricity consumption in Thailand, according to the Ministry of Energy.[32]

31 UNICEF. Heat Waves and How They Impact Children.

32 *The Nation*. 2024. April 29 Sees Record Electricity Usage in Thailand as the Heat Bites. 5 May.

Extreme heat also impacts the availability and integrity of water and wastewater systems for urban use and irrigation, and affects urban infrastructure through material degradation, thermal stress, and amplified heat, thereby affecting livability, economic productivity, and resilience. The hospitality industry in Thailand is facing increasing operating costs due to the need for enhanced cooling systems. Energy consumption in hotels has surged, driven by air-conditioning demand, putting financial pressure on small and medium-sized businesses. Additionally, water shortages caused by prolonged droughts in areas with a high concentration of tourists, such as Phuket and Koh Samui, further strain infrastructure and reduce the quality of guest experiences.

Tourism infrastructure is also being tested by increasing temperatures. Airports, public transportation systems, and even roads are not always designed to withstand prolonged periods of extreme heat. Bangkok's elevated transit system, for example, has faced heat-related operational delays, while ferry operators in southern Thailand have experienced a decline in passenger numbers during heat waves because of safety concerns and discomfort. Without proactive climate-proofing, these logistical disruptions could discourage repeat visitors and reduce tourist satisfaction.

Biodiversity

The increase in the intensity, frequency, and duration of heat waves is causing significant heat stress in all living organisms, leading to habitat loss and altered species distribution. Heat stress has multiple negative effects on plants, as it impacts photosynthesis, respiration, growth, development, reproduction, and susceptibility to contamination with pathogens and decrease the nutritional value of cereal grains, thus threatening agriculture and rainforest worldwide.[33] In recent years, Thailand has experienced considerable biodiversity loss through forest wildfires caused by excessive heat conditions.

Heat stress affects animals resulting in physiological and behavioral changes such as reduced calorie intake, increased water consumption, and lower reproduction rates and growth. Fish find it difficult to breathe in warmer waters because of reduced oxygen levels, leading to increased stress, lower reproductive rates, and higher mortality. Marine ecosystems face challenges—e.g., the Gulf of Thailand's coral reefs, which are highly sensitive to temperature changes, experience coral bleaching, leading to disruptions in marine biodiversity and threatening the livelihood of fishing communities. Warmer nutrient-rich waters are ideal for harmful algal blooms to thrive and release toxins that threaten marine life, contaminate seafood, and pose serious risks to human health.

Addressing these alarming trends is critical, particularly for Thailand and the rest of Southeast Asia, where rising heat poses severe multidimensional challenges, leading to health deterioration, economic strain, infrastructure vulnerability, and ecological decline.

[33] V. Sampath et al. 2023. Acute and Chronic Impacts of Heat Stress on Planetary Health. *Allergy*. 78 (8). pp. 2109–2120.

CHALLENGES TO SCALING UP HEAT STRESS RESILIENCE

Several barriers hinder efforts to scale up adaptation and mitigation activities to quickly meet the challenges of a heating planet. These barriers, illustrated in Figure 6, fall under three broad areas:

(i) lack of awareness at the policy and institutional levels,

(ii) fragmented project development approaches, and

(iii) rising financing gaps.

Figure 6: Key Barriers to Strengthening Heat Resilience

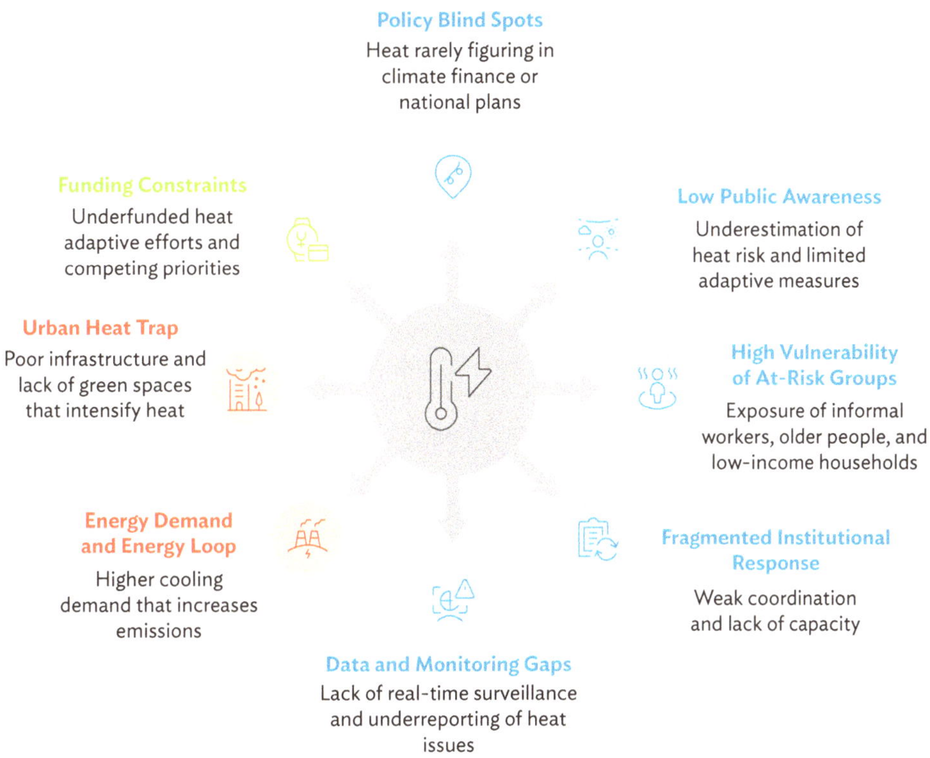

Source: Authors.

Lack of Awareness at the Policy and Institutional Levels

Policy Blind Spots

Extreme heat is among the fastest-growing climate risks in Asia and the Pacific. Yet, unlike floods and storms, extreme heat rarely generates visible infrastructure damage or post-disaster reconstruction needs. Consequently, it is underrepresented in planning systems and public expenditure frameworks that prioritize asset protection and recovery. Heat impacts are distributed across health, labor productivity, urban services, and social protection, with no single implementing agency responsible for comprehensive risk management. This results in a persistent gap between rising risk and policy attention, particularly within investment-oriented climate finance mechanisms.

Low Public Awareness

Heat stress risks are often poorly recognized and frequently underestimated by both the public and the policymakers who, particularly in rural areas, are unaware of the severe health consequences of heat stress. This lack of awareness delays the adoption of adaptive and preventive measures.[34] Insufficient outreach to vulnerable groups, such as elder residents and outdoor laborers, leaves them unprepared to mitigate heat stress. Public health campaigns often fail to reach rural communities effectively because of resource limitation barriers.[35]

High Vulnerability of At-Risk Groups

Marginalized communities facing heightened exposure to extreme heat often lack the knowledge and resources to adapt effectively, such as access to air conditioning, adequate housing insulation, or health care. High-risk sectors such as agriculture and construction, where workers are frequently exposed to outdoor conditions, often lack the required financial or infrastructure support. This not only leads to economic losses but also deepens existing social inequalities, as low-income and informal workers bear the brunt of the impact while having the slightest means to protect themselves.

Fragmented Institutional Response

Dealing with heat stress requires a multisector approach missing in most governments. In the absence of a steering or leading group or institution to spearhead heat stress programs, only fragmented efforts can be made on a sector-specific approach—e.g., through a specific ministry such as environment, labor, energy, and urban planning. There is a shortage of professional urban planners and environmental health experts, and there are no dedicated heat officers, which in turn limits effective response strategies. Thailand's health care infrastructure, for instance, is not yet systematically prepared for heat-related illnesses; moreover, there is an absence of heat–health action planning. Local and urban governments lack capacity, technical knowledge, or access to funding to implement targeted heat resilience actions. The lack of mainstreaming policy responses inhibits urgent systemic action.

Data and Monitoring Gaps

Inadequate heat stress data including localized temperature trends, occupational exposure risks, and heat-related health statistics to guide decision-making persists. Limited real-time heat surveillance inhibits the integration of developments as they unfold into early warning systems or planning tools. Heat-related deaths and illnesses are often underreported. These gaps hinder the development of targeted policies and evidence-based interventions.

34 R. S. Kovats and S. Hajat. 2008. Heat Stress and Public Health: A Critical Review. *Annual Review of Public Health*. 29. pp. 41–55.

35 S. D. Arifwidodo and O. Chandrasiri. 2020. Urban Heat Stress and Human Health in Bangkok, Thailand. *Environmental Research*. 185. 109398.

Fragmented Project Development Approaches

Developing infrastructure projects in isolation could be counterproductive to impacting heat, as illustrated for instance by a transport project focusing on urban road development while removing existing green cover or water bodies. Integrating water, transport, and urban development often with environment or nature-based solutions and defining a combined impact on both the sector indicators themselves as well as on indicators for heat stress and air quality metrics are critically important. Such an approach will ensure better project design and planning and better use of nature-based solutions.

Urban Heat Trap

Urban development requires an integrated approach to deal with the issue of the urban heat island effect, a phenomenon where urban areas become significantly warmer than surrounding rural regions, especially at night (Figure 7). This occurs because cities are dominated by heat-absorbing materials such as concrete, asphalt, and densely packed buildings that store heat during the day and release it slowly after sunset. Additionally, urban environments generate waste heat from vehicles, air conditioners, and industrial activities, intensifying the heat effect. The lack of green spaces, poorly ventilated structures, and limited vegetation further reduce natural cooling through shade and evapotranspiration. Consequences include higher energy consumption, worsened air quality and elevated ground-level ozone, public health risks, as well as ecosystem disruptions and greater water stress.

Figure 7: The Urban Heat Island Effect

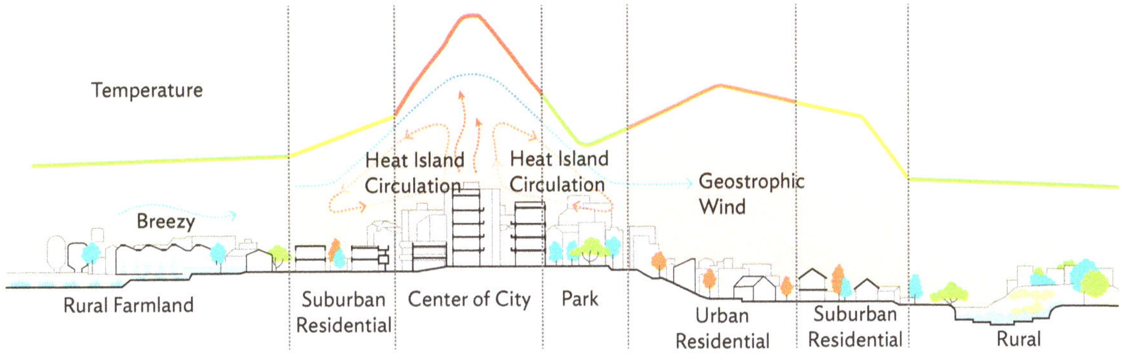

Source: Adapted from Asian Development Bank. 2022. *Beating the Heat: Investing in Pro-Poor Solutions for Urban Resilience*.

Energy Demand and Energy Loop

Similarly, the development of buildings is generally not integrated with the impact heat can have. Buildings are not designed for passive cooling or ventilation, and urban design and zoning codes often neglect thermal comfort and resilience. This results in increased use of air-conditioning that raises energy demand and emissions, contributing to increased CO_2 and straining the grid. A shift to clean cooling technologies and energy-efficient practices is required.

Rising Financing Gaps

Funding Constraints

Country financing needs especially to deal with the impacts of a volatile climate and support the development of resilient infrastructure are rising, and so does the financing gap that often reaches 50% or more of the need.

With financing needs spread across numerous sectors, new and emerging needs such as heat stress or clean air infrastructure are often left out. For instance, Thailand alone needs $22 billion–$28 billion yearly in climate investments in the period 2030–2050 to meet its nationally determined contribution and net-zero targets. Current financial flows fall short by 50%, creating an annual investment gap of $11 billion–$17 billion, according to a recent ADB report.[36]

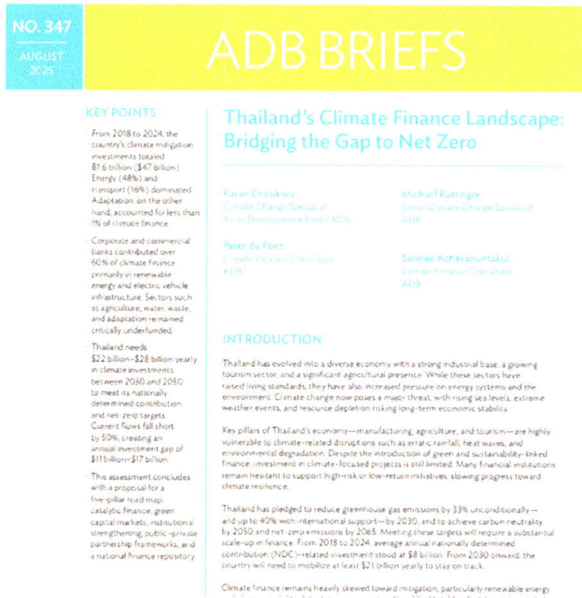

NO. 347
AUGUST 2025

ADB BRIEFS

KEY POINTS

- From 2018 to 2024, the country's climate mitigation investments totaled B1.6 trillion ($47 billion). Energy (48%) and transport (16%) dominated. Adaptation, on the other hand, accounted for less than 1% of climate finance.
- Corporate and commercial banks contributed over 60% of climate finance primarily in renewable energy and electric vehicle infrastructure. Sectors such as agriculture, water, waste, and adaptation remained critically underfunded.
- Thailand needs $22 billion–$28 billion yearly in climate investments between 2030 and 2050 to meet its nationally determined contribution and net-zero targets. Current flows fall short by 50%, creating an annual investment gap of $11 billion–$17 billion.
- This assessment concludes with a proposal for a five-pillar road map: catalytic finance, green capital markets, institutional strengthening, public–private partnership frameworks, and a national finance repository.

Thailand's Climate Finance Landscape: Bridging the Gap to Net Zero

Karan Chouksey
Climate Change Specialist
Asian Development Bank (ADB)

Michael Rattinger
Senior Climate Change Specialist
ADB

Peter du Pont
Climate Finance Consultant
ADB

Sarinee Achavanuntakul
Climate Finance Consultant
ADB

INTRODUCTION

Thailand has evolved into a diverse economy with a strong industrial base, a growing tourism sector, and a significant agricultural presence. While these sectors have raised living standards, they have also increased pressure on energy systems and the environment. Climate change now poses a major threat, with rising sea levels, extreme weather events, and resource depletion risking long-term economic stability.

Key pillars of Thailand's economy—manufacturing, agriculture, and tourism—are highly vulnerable to climate-related disruptions such as erratic rainfall, heat waves, and environmental degradation. Despite the introduction of green and sustainability-linked finance, investment in climate-focused projects is still limited. Many financial institutions remain hesitant to support high-risk or low-return initiatives, slowing progress toward climate resilience.

Thailand has pledged to reduce greenhouse gas emissions by 33% unconditionally—and up to 40% with international support—by 2030, and to achieve carbon neutrality by 2050 and net-zero emissions by 2065. Meeting these targets will require a substantial scale-up in finance. From 2018 to 2024, average annual nationally determined contribution (NDC)-related investment stood at $8 billion. From 2030 onward, the country will need to mobilize at least $21 billion yearly to stay on track.

Climate finance remains heavily skewed toward mitigation, particularly renewable energy

Asian Development Bank (ADB) report analyzing Thailand's climate finance landscape. (Screen grab from ADB website).

Meanwhile, the World Economic Forum (WEF) has reported that, in 2015–2021, less than 1% of international development funding and only 2% of international public climate finance targeted air pollution action.[37]

Private capital is especially hard to mobilize where underlying projects are considered unbankable or unable to provide a sufficient return given the perceived risk. Funding for innovations such as energy-efficient cooling technologies or heat-resistant infrastructure is minimal. Competing financial priorities leave heat mitigation underfunded. Heat stress is not prioritized in climate finance. Climate finance for heat-specific interventions (e.g., cooling centers, green roofs) remains scarce. Blended finance mechanisms and private sector engagement are still in the early stages.

There are promising air quality programs being developed by international development agencies that can also be leveraged to support heat stress actions. These include an initiative by the Multilateral Investment Guarantee Agency, with an announced $1 billion to derisk air quality projects and encourage more investment in clean air; and ADB's Asia Clean Blue Skies program, which aims to scale up investments and support capacity building in air quality improvement and pollution control in developing countries in Asia and the Pacific. However, given the scale of the financing needs, much more must be done to catalyze the flow of finances from all sources, especially the private sector and capital markets.

36 K. Chouksey et al. 2025. Thailand's Climate Finance Landscape: Bridging the Gap to Net Zero. *ADB Briefs*. No. 347. ADB.
37 M. G. Bathan-Baterina and A. Mehra. 2024. Five Ways Development Finance Institutions Can Accelerate Air Clean Air Financing. *WEF*. 10 July.

EMERGING SOLUTIONS FOR HEAT RESILIENCE

A Tool Kit of Solutions

A wide range of solutions—technical, institutional, policy—are increasingly being implemented by governments, cities, and institutions to address the growing impacts of heat stress and can be examples to build on and scale up globally. These interventions are critical for reducing vulnerability, ensuring public health, and enhancing resilience in the face of rising temperatures. As shown in Table 4, the interventions involve behavioral, regulatory, and technical strategies as well as nature-based solutions that offer a comprehensive framework for mitigating heat stress and adapting to a warmer planet. Together, they represent a tool kit of practical and scalable interventions that can be curated to local requirements.

Table 4: Strategies for Mitigating Heat Stress

Behavioral Solutions Individuals modify behavior to reduce heat exposure	**Regulatory Solutions** Government implements policies for community heat resilience	**Technical Solutions** Technical interventions help reduce ambient temperature	**Nature-Based Solutions** Application of natural blue and green solutions reduces heat stress impact
• Adjusting work, leisure, school hours • Wearing appropriate clothing • Public information campaigns promoting hydration awareness • Using fans, shades, etc. • Community education • Heat risk mapping • Public cooling centers • Urban gardening, tree planting • Heat wave preparedness campaigns • Cross ventilation, curtain use • Water provisioning	• Mandatory heat action plans • Occupational safety standards • Urban planning regulations • Urban heat monitoring networks • Use of GIS and remote sensing to identify heat spots • Inclusion of heat risk in national climate adaptation strategies • Revising building codes • Enforcement of zoning laws • Integration of heat alerts into disaster management protocols • Subsidies, incentives for heat-resilient buildings and cooling technologies • Health and labor policies	• Building design (thermal mass) • Urban, spatial planning • Reflective building materials • Urban planning regulations • Public cooling shelters • Early warning systems • Retrofitting homes with passive cooling design • IOT-enabled heat monitoring sensors • District cooling systems • City-based heat information apps • Early warning systems • Leveraging AI climate and health data integration	• Urban greening and tree canopy expansion • Green roofs and walls • Outdoor shade • Permeable pavements • Blue, green infrastructure • Smart irrigation systems • Wetlands • Agroforestry and shade cropping • Climate-resilient crops

AI = artificial intelligence, GIS = geographic information system, IOT = Internet of Things.
Source: Authors.

Behavioral Solutions

Behavioral solutions focus on changing human actions to reduce heat exposure and its effects. In Thailand, these solutions include promoting practices such as wearing lightweight and breathable clothing, staying hydrated, reducing outdoor activities during peak heat hours, and increasing awareness about recognizing early signs of heat stress. Public education campaigns and community training can help individuals and workers adopt safer routines, especially in agriculture, construction, and other outdoor sectors where exposure is greatest.

Regulatory Solutions

Regulatory solutions involve government policies and labor protections to safeguard vulnerable populations. This means enforcing workplace heat safety standards, mandating rest breaks for outdoor workers, and setting maximum allowable working hours in high-temperature environments. Regulations could also include urban planning guidelines to reduce urban heat island effects and building codes requiring heat-resilient designs. Strong enforcement mechanisms and collaboration with industries are important for ensuring compliance and protecting both formal and informal workers.

Technical Solutions

Technical solutions rely on innovation and technology to mitigate heat exposure. In Thailand, these include the use of cooling technologies such as misting systems, improved ventilation in buildings, reflective roofing materials, and wearable cooling gear for workers. Smart city technologies like heat mapping can identify hotspots and guide targeted interventions. Expanding access to affordable cooling appliances, such as energy-efficient fans and air-conditioning, can also significantly reduce heat-related health risks, particularly in vulnerable communities.

Nature-Based Solutions

Nature-based solutions harness ecosystems to reduce heat stress and create more livable environments. Expanding urban green spaces, planting shade-providing trees along streets, and restoring wetlands can help lower local temperatures and improve air quality. Rooftop gardens, vertical greenery, and community parks contribute to natural cooling while enhancing biodiversity and resilience. These approaches not only mitigate heat but also provide co-benefits such as flood management, carbon absorption, and improved quality of life in cities. Box 2 highlights the notable work done by the International Union for Conservation of Nature to improve air quality and mitigate heat stress through nature-based solutions.

Box 2: The International Union for Conservation of Nature and Nature-Based Solutions for Air and Heat

The International Union for Conservation of Nature (IUCN) has noted the intertwining of the air and heat crisis for most of the fast-growing cities of Southeast Asia as they expand. Megacities and secondary cities lose their green spaces as they expand, and the spread of impervious construction material intensifies the urban heat island effect, trapping heat and pollutants close to the ground. Further, the **IUCN's Asia Nature-Based Solutions Programme** has shown that many Asian cities are now facing not just extreme heat but also frequent flooding as unplanned urban growth and changes in the climate compound the challenges. This underscores the urgent need to incorporate **ecosystem-based approaches** into city planning and infrastructure investment.[a]

The work of IUCN on nature-based solutions (NBS) highlights the cooling power of urban green and blue spaces—from street trees, parks, and wetlands to natural water courses, retention ponds, and restored riparian corridors. Planning guidance initially developed for Mediterranean cities demonstrates that well-designed NBS can lower city temperatures through shade, evapotranspiration, and improved airflow. At the same time, these systems help remove air pollutants,

continued on next page

Box 2 *continued*

store carbon, and improve overall air quality.[b] IUCN's guidance on building **urban ecological networks** across the Association of Southeast Asian Nations (ASEAN) emphasizes the importance of connecting green corridors, riverfront buffers, and biodiversity-rich spaces to cool neighborhoods and make public areas more comfortable. Prioritizing tree planting along roads and transit routes, expanding parks in low-income communities, and restoring urban waterways can generate cooler microclimates—particularly benefiting outdoor workers, informal settlements, and vulnerable groups.[c]

The **IUCN Global Standard for Nature-Based Solutions** also offers a framework to ensure that such interventions are not only effective but also climate-resilient, socially inclusive, and biodiversity-positive. It calls for a sound understanding of local problems, strong stakeholder engagement, transparent trade-off management, and sustainable financing for long-term upkeep. NBS should not be seen as a beautification option but an essential infrastructure for climate resilience and public health.

NBS = nature-based solution.

[a] IUCN. Nature-Based Solutions for Climate Adaptation and Mitigation.

[b] IUCN. 2021. *Planning and Delivering Nature-Based Solutions in Mediterranean Cities: First Assessment of the IUCN NbS Global Standard in Mediterranean Urban Areas—Summary for Policy-Makers*.

[c] M. Sharma, M. K. S. Pasha, and O. Chassot. 2025. *Building Urban Ecological Networks for Conservation in the ASEAN Region: A Discussion Paper on Increasing Connectivity Amongst Urban Protected Areas, OECMs, and Green and Blue Spaces*.

Source: International Union for Conservation of Nature Thailand.

Snapshots of Emerging Solutions

A growing number of case studies are emerging, highlighting solutions and lessons that can be built upon in the design and implementation of heat resilience interventions. Some of these cases are outlined below, under the key areas of policy, capacity, and projects.

National Planning Approaches

- ***Singapore: Singapore Green Plan 2030—A Whole-of-Nation Approach to Sustainable, Climate-Resilient Development.*** The Singapore Green Plan 2030 outlines a whole-of-nation approach to the sustainable and climate-resilient development of the city-state into a city in nature. It sets out decarbonization and climate adaptation strategies on clean energy, green transport, sustainable infrastructure, urban cooling, strengthening of coastal and flood defenses, food security, zero waste, and a circular economy. The Ministry of Sustainability and the Environment launched the S$50 million SG Eco Fund to finance ground-up projects that support sustainable initiatives by the people, private, and public (3P) sectors. The SG Eco Fund prioritizes communities and fosters partnerships for environmental sustainability. Additionally, Singapore employs targeted funds and bonds tailored for specific industries and groups to bolster resilience programs.[38]
- ***Kyrgyz Republic: Heat Wave Early Action Protocol.*** The Kyrgyz Republic is extremely vulnerable to climatic changes as it has high exposure to both natural and human-induced hazards. Most pessimistic estimates place the projected average temperature increase in the country at 8°C during the century.[39] To prepare for this eventuality, key ministries and agencies, as well as local municipalities and community leaders, collaborated with the Red Crescent Society of Kyrgyzstan, the German Red Cross, and other humanitarian organizations to prepare an early action protocol for heat waves. The protocol targets vulnerable families below the poverty line

38 Government of Singapore, Singapore Green Plan 2030. Funding for Communities, Schools, Businesses.

39 UNICEF Kyrgyzstan. 2017. Kyrgyzstan is One of the Most Vulnerable Countries to Climate Change in Central Asia. *Press release*. 22 March.

as well as older people, and introduces forecast-based financing to deliver timely cash, cooling solutions, public alerts, educational material, and public awareness support to at-risk populations.[40] With a counterpart for cold waves, the protocol provides a model for harnessing anticipatory action—guided by weather data and local engagement—to ensure low-cost, life-saving interventions.

City Planning Approaches

- ***Ahmedabad, Gujarat, India: Heat Action Plan.*** Ahmedabad pioneered South Asia's Heat Action Plan, combining public awareness campaigns targeting low-income and marginalized groups, early warning systems, and capacity building for health care professionals, with interagency coordination, as key interventions. Its main innovation is the cool roof program that provided free reflective paint and lime coatings for 3,000 low-income households, promoting heat resilience in vulnerable communities.[41] The initiative demonstrated a scalable low-cost solution, as voluntary adoption of cool roofs followed the pilot phase. It also illustrated the importance of collaborative funding models, with public–private partnerships playing a key role and local businesses providing free materials for the program.
- ***Greater Sydney, Australia: Heat Smart City Plan.*** The Greater Sydney Heat Smart City Plan, 2025–2030 is a collaborative initiative involving a broad partnership of government, industry, and community organizations that aims to tackle heat-related risks and improve urban resilience.[42] Its six pillars—Heat Smart Decisions, Heat Smart Places for People, Heat Smart Economies, Heat Smart Infrastructure, Heat Wave Ready, and Heat Smart Research—integrate heat resilience into urban planning, infrastructure, governance, risk monitoring, and economy-wide decision-making. The plan is funded through national and state disaster risk reduction programs and provides a long-term blueprint for reducing vulnerability across systems.
- ***Amsterdam, Netherlands: Heat Plan 2024.*** Amsterdam's Heat Plan 2024, prepared by GGD Amsterdam, the city's Public Health Service, in collaboration with emergency and relief agencies, centers on protecting vulnerable populations. The plan is activated when temperatures exceed 27°C for 4 consecutive days.[43] The city's targeted interventions for outdoor workers, older people, children, and individuals with existing health conditions are a strong example of integrating public health into municipal climate adaptation.
- ***Mexico City, Mexico: Green Challenge (Reto Verde).*** Mexico City's Green Challenge (Reto Verde), initiated in 2019 by the Ministry of Environment's Green Infrastructure Directorate in collaboration with other sectors of the city government, is a greening strategy with important lessons for integrating urban development, air quality, and heat stress impacts. It aims to increase nature and biodiversity in public spaces, reduce the impact of climate volatility, prevent soil erosion, and improve air and water quality across the entire city. The original goal to plant more than 10 million trees and other plants by the end of 2020 was exceeded by an ambitious collaborative urban greening program of planting 44.2 million trees, shrubs, and ground covers since 2019, with a revised target of 50 million trees and other plants by 2024.[44] Urban green space consequently rose from 14.5 square meters per person in 2017 to 19.4 square meters per person in 2020. The program established 16 vertical gardens and over 1,000 pollinator gardens and contributed to integrated biodiversity restoration by rehabilitating natural protected areas and creating wetlands to improve ecosystem services and mitigate heat. The Green Challenge program relies on the combined efforts and resources from city government departments, municipal governments, civil society, and private companies.[45] Collaborative funding and resource pooling have been critical for achieving the program's ambitious goals.

[40] German Red Cross and Red Crescent Society of Kyrgyzstan. 2019. Forecast-Based Financing: Kyrgyzstan.
[41] UNEP. 2021. *Beating the Heat: A Sustainable Cooling Handbook for Cities*.
[42] Western Sydney Regional Organisation of Councils. Greater Sydney Heat Smart City Plan.
[43] Municipality of Amsterdam. 2024. *Amsterdam Heat Plan 2024: A Healthy Social City Even During Heat Waves* (in Dutch).
[44] *Mexico Business News*. 2024. Mexico City's Green Urban Efforts. 20 February.
[45] Urban Sustainability Exchange. Green Challenge.

- ***Ferozepur, Punjab, India: District Heat Action Plan (2024).*** This is a proactive district-level response developed by the District Disaster Management Authority, in collaboration with multiple government and civil society agencies, to protect both human and animal population during extreme heat events.[46] The plan defines heat wave criteria based on temperature thresholds and deviations from normal—e.g., for plains, heat waves are declared when maximum temperatures reach or exceed 45°C. It involves alerting at-risk populations, implementing preparedness measures, and mobilizing departments and communities to address health impacts during extreme heat events. The plan is funded through a combination of disaster management funds, including the National and State Disaster Mitigation Funds, as well as city and state budgets. It demonstrates the potential of local governments and decentralized governance in driving heat wave response effectiveness.
- ***Kigali, Rwanda: Urban Wetland Restoration.*** Kigali's restoration of the Gikondo and Nyabugogo wetlands offers a practical nature-based approach to cooling in a rapidly urbanizing city. By rehabilitating these degraded wetlands and integrating them into city planning, Kigali has reduced local surface temperatures in dense neighborhoods while improving flood control and biodiversity. The initiative shows how protecting natural ecosystems can deliver low-cost cooling benefits and strengthen overall urban climate resilience.

Technological Advancement Examples

- ***Japan: Heat-Reducing Materials.*** Special fabrics have been developed to block heat and these have been used increasingly in Japan for heat-blocking parasols, heat-releasing clothing, and agricultural sheets.[47] An example is the "Summer Shield" special fabric developed by Toray, which blocks more than 99% of ultraviolet rays and can provide a heat-blocking effect of −4°C and is now recommended to avoid heatstrokes. The "Kuchofuku" air-conditioned uniform is another interesting example, featuring a built-in fan in the jacket, which blows air to reduce perspiration and replaces warm air with cool outside air to keep the body cool. Agriculture sector innovations such as special sheets and nets for greenhouses, which can lower temperatures sometimes by −2°C, help farmers avoid heat stress.
- ***Indonesia: Cool Roofs.*** An emerging model for heat reduction in buildings is from Indonesia.[48] The model uses "cool" roofs, which have a special coating to reflect sunlight and reduce the amount of heat absorbed by a building, thus lowering indoor temperatures. The model was introduced under the Million Cool Roofs Challenge in the city of Tangerang, an industrial hub near Jakarta and home to the Soekarno-Hatta International Airport, as well as on the roofs of more than 40 public and community buildings. It is reported that, in one industrial building, the indoor temperature dropped by 10.6°C, from 40.0°C to 29.4°C, following the cool roof adoption.
- ***Barcelona, Spain: Cool Schools.*** Barcelona's "climate shelters" retrofit public schools with shading, greenery, and reflective materials, lowering temperatures by 4°C–5°C. Schools double as community cooling centers during heat waves, offering a child-centered and community-wide model for heat resilience.
- ***Other Technological Innovations.*** The Institute of Development Studies in the United Kingdom[49] has identified other technological advancements in numerous areas that could provide solutions for scaling up heat resilience. These include (i) trees in croplands alleviating heat stress among agricultural workers in Tanzania; (ii) cool roofs in New Zealand; (iii) coproduced, complex heat adaptation interventions to reduce heat impacts on pregnant women and newborns in Southern Africa (an intervention development and feasibility evaluation study, led by Wits Health Consortium, South Africa); and (iv) housing modifications for indoor thermal comfort in Kenya.

---

[46] District Disaster Management Authority, Ferozepur. 2024. *Heat Wave Action Plan, Year 2024*. District Ferozepur, Punjab.

[47] N. Tochibayashi and M. Ota. 2024. Cooling Japan: How Innovative Materials Are Tackling Heat Waves. *World Economic Forum*. 31 July.

[48] N. C. Chin and J. Wu. 2023. Save Money, Be Cool: Indonesian Project Shows How 'Cool' Roofs Can Help Asia Beat the Heat. *Channel News Asia*. 24 May.

[49] Institute of Development Studies. 2025. Finding Innovative Ways to Adapt to Our Heating Planet. 10 January.

City Project Examples

- ***Seoul, Republic of Korea.*** An effort to restore the Cheonggyecheon stream that runs through the city replaced 5.8 kilometers of elevated expressway covering the stream with a mixed-use waterfront corridor. The waterfront corridor decreased the temperature in the area by 3.3°C–5.9°C compared to a parallel road a few blocks away.
- ***Medellín, Colombia.*** Green corridors were created that follow and restore the geography of the area prior to recent development. In 2016–2019, the city created 36 corridors, 18 along major roads and 18 along waterways, covering over 36 hectares. The areas with green corridors have already seen temperature reductions of up to 4°C.
- ***Paris, France.*** Paris is home of the first and largest district cooling system in Europe. When the water temperature in the Seine River that cuts through the city is below 8°C, this water is used to provide "free cooling."
- ***Toronto, Canada.*** The municipal government implemented the largest lake-source cooling system in the world. Commissioned in 2004, Enwave's 264-megawatt deep lake water cooling system uses Lake Ontario's cold water as a renewable energy source.
- ***Guangzhou, People's Republic of China.*** The municipal government adopted regional centralized cooling as part of a green and environment-friendly modern urban center in the Pearl River New City development. The resulting local temperature in the core area of Zhujiang New Town was 2°C–3°C lower compared to the temperature achieved through distributed cooling systems.

EMERGING FINANCE SOLUTIONS

Broadening the Sources of Finance: The Need for Private Capital

Given the major pressures on government balance sheets everywhere, the key to scaling up finance into a larger pipeline of heat-impacting projects is the private sector. This is even more true at present with volatile economies and competing needs to address, including food security, climate-resilient infrastructure, and much more. According to the Asian Development Bank (ADB), developing Asia alone needs more than $1.7 trillion a year until 2030 to meet its infrastructure needs,[50] and this will require private capital to step in. However, ensuring the bankability of these projects will be critical for attracting such funds. This, therefore, suggests that a much better use of scarce government and development fund budgets is required to reduce the risk of such projects and to enhance their *perceived* ability to generate the returns or profits needed to assure repayment of finance raised from private sector investors —an approach often referred to as "derisking."

Available Public Budgets for Derisking

Given the crosscutting impacts of extreme heat, several government budget streams can be better leveraged for this derisking function, for instance, from ministries or agencies of health, energy, agriculture, transport, urban, and environment. Consistent government budget allocations—enshrined in 5-year plans—can ensure a scaling up of implementation in phases, particularly for vulnerable areas. These government budget allocations for heat can then be further enhanced with funds from global development, philanthropic, and climate-focused funds. Figure 8 shows several financing sources that government and others can tap for project heat derisking.

Financing Instruments and Approaches

National Derisking, Incentivizing, or Blended Heat Funds

Structuring and funding national funds or facilities to provide funding on advantageous terms is a green financing approach that can incentivize investments in heat-impact projects. Efforts can build on the experience of examples already in place, including the SDG Indonesia One: Green Finance Facility (Figure 9) and the Shandong Green Development Fund, both supported by ADB. Another emerging example is the Monetary Authority of Singapore's blended finance partnership under Singapore's Financing Asia's Transition Partnership (FAST-P) initiative. This blended finance partnership is collating over $500 million in capital from global and regional private, public, and philanthropic institutions for use in derisking green and sustainable infrastructure opportunities across Southeast Asia and South Asia.

---

50 ADB. 2017. *Meeting Asia's Infrastructure Needs*.

Figure 8: Available Financial Resources for Project Heat Derisking

Source: Authors.

Figure 9: SDG Indonesia One: Green Finance Facility

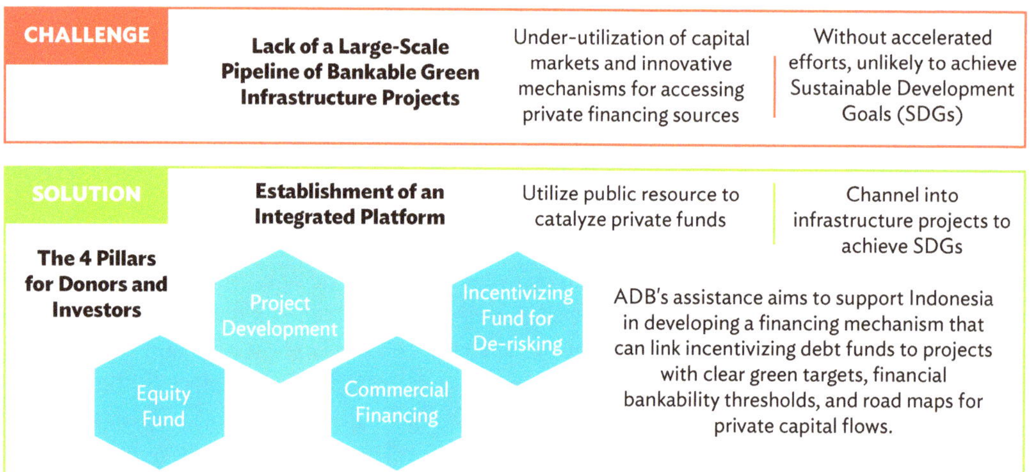

Source: Brochure produced by the Asian Development Bank in 2022 on SDG Indonesia One: Green Finance Facility.

National funds, as exemplified by the United Kingdom's Green Heat Network Fund (Box 3), could similarly be established to catalyze heat impact-related funds. Funds provision may be linked to a few conditions, including (i) measurable heat impact-related indicators, (ii) a percentage of private capital crowding into the project, and (iii) enhancement of project bankability. Climate finance facilities such as the Green Climate Fund, Global Environment Facility, Adaptation Fund, and various forms of bilateral aid could be explored to develop national heat funds.

Box 3: The United Kingdom's Green Heat Network Fund

The Green Heat Network Fund (GHNF) is a dedicated fund established by the United Kingdom (UK) government in 2022 initially at £288 million to finance projects that aim to develop low- and zero-carbon heat (and cooling) networks. Funding comes in the form of capital grants for both the commercialization and construction of new low- and zero-carbon heat networks, along with the retrofitting and expansion of existing ones.

The GHNF supports the UK government's Heat Network Transformation Programme, which addresses some of the challenges in decarbonizing the UK's heat sector. The UK estimates that, by 2050, around 18% of the country's heat will need to come from heat networks if it is to meet its carbon targets cost effectively. Heat networks are an established method of distributing heat that can utilize otherwise wasted energy, provide grid balancing services in an increasingly electrified heat market, and offer low-carbon heat at competitive prices to households and businesses alike.

Source: Government of the United Kingdom. 2022. Green Heat Network Fund Scheme Overview. Department of Energy Security and Net Zero and Triple Point Heat Networks.

The Netherlands Heat Fund[51] is an interesting model of a government–private finance collaboration fund that has emerged recently. This fund aims to accelerate household-level energy transition by providing loans to private individuals with interest rates linked to income thresholds and interest-free loans available to some households. All funding is for household investment in energy-saving measures. Eligible energy-saving measures include high efficiency boilers, heat recovery systems, heat pumps, solar water heaters, etc. While the government established the initial fund, commercial banks have also contributed. As of 2025, the International Netherlands Group had contributed a total of €600 million in funding and the Rabobank a total of €550 million in financing and guarantees.[52] As of 2024, the fund granted a record €426 million in loans and in total has benefited nearly 30,000 households.

Emergency Response and Disaster Risk Reduction Funding Programs

Emergency response and disaster risk reduction funding programs include national and internationally funded programs provided both in response to an extreme event (wildfires), as well as to build resilience and reduce the risk of heat-related impacts—e.g., through the development of early warning systems and community-based disaster preparedness plans.

51 Nationaal Warmtefonds. Borrow Money for Energy-Efficient Living: Green Light for Your Sustainability.

52 International Netherlands Group. 2025. ING Netherlands Makes Another €100 Million Available for the National Heat Fund. 9 September.

Capital Market Instruments

Green and sustainability-linked bonds have been shown to crowd in a large amount of finance from domestic or global investors, if structured well. These bonds link the usage of funds with clear climate or environment objectives and well-developed, measurable indicators. They have seen widespread adoption in the European Union and are gaining momentum in Asia, including Thailand, with governments successfully issuing bonds to support green development.

- ***Thailand's Sustainability-Linked Bond.*** Issued in 2024, this is the first sovereign sustainability-linked bond in Asia. It raised about $880 million and was almost three times oversubscribed.[53] Unlike traditional green bonds that focus on use-of-proceeds, this sustainability-linked bond ties financial outcomes directly to measurable national climate targets, marking a strategic evolution in Thailand's sustainable finance landscape. The issuance demonstrates how policy innovation, supported by international technical collaboration, including from ADB, can help overcome systemic limitations and build the case for future performance-based instruments aligned with climate and urban resilience goals.

Thailand's Sustainability-Linked Bond. Representatives from the Asian Development Bank; the Public Debt Management Office under the Ministry of Finance, Government of Thailand; and other partners attend a road show to launch Thailand's sovereign sustainability-linked bond, the first-of-its-kind in Southeast Asia, December 2024 (photo by ADB Thailand Resident Mission).

- ***City of Paris Climate Bond.*** This is a €300 million bond issued in 2015 when the city hosted the 21st Conference of the Parties of the United Nations Framework Convention on impact due to changes in the climate.[54] With a maturity date of May 2031, the bond is targeted at financing climate and energy projects. Paris had already identified the growing threat from rising average daily temperatures, with the increasing frequency of hot, very hot, and extremely hot days and heat waves. A 2023 City of Paris report projected that by 2085, the city could face persistent heat waves with more than 34 days a year exceeding an average daily temperature of 30°C, and peak summer temperatures reaching around 50°C.[55] The devastating 2003 heat wave, leading to almost 15,000 heat-related deaths, underscored the city's vulnerability to extreme heat. In response, Paris developed a heat wave plan, introducing measures such as expanding green spaces and rooftop gardens, increasing urban parks, and use

53 ADB. 2024. ADB Supports Asia's First Sustainability-Linked Bond in Thailand. *News release*. 9 December.

54 Climate ADAPT. 2023. Climate Bond Financing Urban Adaptation Actions to Reduce Heat Stress in Paris, France. Case Studies.

55 City of Paris. 2023. *Paris at 50°C: A Report by the Paris Council Information and Evaluation Committee* (in French).

of water features for cooling. The Paris Climate Bond was able to raise considerable funding from private and institutional investors, 20% of which has been earmarked for adaptation, including projects to reduce heat stress in the city—e.g., through targeted planting of 20,000 new trees on streets, school playgrounds, gymnasiums, etc. of inner Paris; and the creation of 30 hectares of new parks in urban renewal zones and existing neighborhoods of Paris. Since 2017, the climate bond has operated on a broader sustainability bond framework, financing a wide range of projects dealing with sustainability.

- ***Pooled Municipal Bonds***. Cities globally are increasingly issuing climate adaptation bonds, resilience bonds, and pooled municipal funds to finance cooling corridors, reflective pavements, water retention systems, and green roofs. Instruments such as municipal pooled financing (as used in India), land value capture, and climate impact-linked infrastructure charges can help cities in Southeast Asia generate the needed local revenue streams for heat resilience bonds to successfully attract the required investment flows.

Revenue Enhancement Approaches

- ***Carbon Pricing and Heat Resilience Allocations***. To promote the bankability of projects, governments could consider setting up mechanisms to raise additional or alternative revenue streams to internalize the hidden costs (negative externalities, e.g., pollution) of economic activities or benefits from projects (positive externalities, e.g., savings in public health expenditure due to heat adaptation and mitigation projects). These revenues can be plowed back to projects to make them more sustainable, as well as scale up coverage and impact. Carbon pricing or carbon taxes in projects with both measurable heat reduction and greenhouse reduction indicators could amplify the potential additional revenues.
- ***Cooling-as-a-Service Models***. The transition to efficient cooling in Thailand—through district cooling, high-efficiency air-conditioning, passive design, and cool roofs—requires innovative financing models that reduce upfront costs and attract private investment. Cooling-as-a-Service, where building owners pay for cooling as a utility while service operators finance and operate such high-efficiency systems, can help overcome capital barriers and accelerate adoption in urban areas. Complementary instruments such as on-bill financing, cooling performance contracts, leasing models, and energy-as-a-service platforms, already used in regions such as the Middle East and Singapore, could be mainstreamed into Thailand's urban development and building regulations to mobilize private financing at scale.

Risk-Sharing Instruments

- ***Parametric Insurance***. Index-based insurance products, triggered by extreme heat events (wildfires, droughts, heat-induced illnesses, heat-related workplace accidents, etc.), protect farmers and outdoor workers, and are widely used to transfer risk and for financial protection. Parametric insurance plans trigger a payment once an established temperature, humidity, or rainfall threshold has been crossed. Loss of earnings due to extreme heat is prevalent in India where the Women's Climate Shock and Insurance and Livelihoods Initiative was launched by the Climate Resilience for All and the Self-Employed Women's Association (India) to support vulnerable women workers.[56]

  Weather-indexed crop insurance has been successfully implemented in several South Asian and Southeast Asian countries, acting as a safeguard and providing timely relief to farmers whose yields have been impacted by extreme heat or dry spells. Unlike indemnity-type insurance policies that require physical field inspections or damage verification, parametric insurance enables quicker payment and reduced administration costs.
- ***Catastrophe Bonds.*** Also called regional risk pools, catastrophe bonds are under the Association of Southeast Asian Nations Disaster Risk Financing and Insurance. These can also be adapted for extreme heat events, though current products focus more on floods and typhoons.

[56] SwissRe. Protecting Vulnerable At-Risk Workers from the Impacts of Extreme Heat—India: Index Insurance for Extreme Heat.

Public–Private Partnerships

Public–private partnerships (PPPs) to mobilize investments into heat-smart urban infrastructure and amenities (e.g., urban greening, cooling centers, reflective materials for buildings) can be structured with climate risk-adjusted returns. Tax breaks, viability gap funding, or low-interest loans can make these projects more attractive. Blended finance models—where risk and investment are shared through the introduction of concessional public funds or other fiscal incentives and philanthropic funding to catalyze private sector investment—can also be explored.

The PPPs are emerging as a powerful tool for climate adaptation and development of green infrastructure. By leveraging the resources, expertise, and innovation of private stakeholders, governments can scale up investments in sustainable solutions that enhance urban resilience. PPPs enable the sharing of risks, costs, and benefits, making large-scale climate adaptation projects financially viable while ensuring long-term sustainability. Developers can be incentivized to include heat stress features in exchange for tax rebates, floor area ratio bonuses, and fast-track permits. Governments and private developers can collaborate on traditional infrastructure using PPPs as well as on other areas such as (i) increasing green spaces through investments in more parks and trees; (ii) scaling up green roofs and green buildings; (iii) replacing traditional heat-absorbent materials with efficient materials (e.g., on pavements) and by retrofitting existing buildings; and (iv) implementing district cooling networks to cool multiple buildings.

A SUGGESTED APPROACH TO HEAT RESILIENCE IN THAILAND

Existing Initiatives

Thailand has implemented a heat wave warning system and issues public health advisories through the Ministry of Public Health's Disease Control Department, particularly during severe heat events, urging high-risk groups—such as children, older people, and individuals with preexisting health conditions—to stay indoors and stay hydrated. Initiatives addressing heat stress around the country include the following:

- ***Educational Heat Action Plans.*** Following a pilot program in Kamphaeng Phet Province, the Ministry of Education introduced nationwide school schedules adjusted for extreme heat.[57]
- ***Green Building Projects.*** Nationally Appropriate Mitigation Actions focus on greening low-income housing and government buildings, indirectly reducing heat stress.[58]
- ***Urban Tree Planting.*** Agoda's initiative supports Bangkok Metropolitan Authority's goal to plant 1 million trees in 4 years.[59]
- ***Public Health Interventions.*** Heat health advisories, mobile health units, and preventive guidelines from local authorities, including recommendations for hydration, clothing, and elder care are being disseminated.
- ***Green Bangkok 2030.*** This program aims to increase green space, tree coverage, and pedestrian-friendly infrastructure in the city, supported by projects like "Big Trees" and new park developments.[60]
- ***Udon Thani Heat-Health Surveillance Pilot.*** This prototype digital surveillance system collects hospital data, heat illnesses, absenteeism trends, and local climate indicators. This mirrors Singapore's dengue surveillance model—but for heat.[61]
- ***Agricultural Resilience***. Programs like the UN's Scaling Up Climate Ambition on Land Use and Agriculture program emphasize climate-resilient agriculture to mitigate heat's impact on the agro-industry.

Scaling Up Action Nationally

While the country has broader climate and health frameworks in place—such as the Climate Change Master Plan (2015–2050), the National Climate Change Adaptation Plan, and various environmental health strategic plans—these primarily address general adaptation, disaster preparedness, and environmental health rather than focused, actionable responses to extreme heat. Similarly, the draft Climate Change Bill includes long-term targets like

57 *Bangkok Post*. 2024. Adapting to a Hotter World. Editorial. 8 May.

58 UN Environment Programme (UNEP). 2018. Building for Green Growth in Thailand. 12 March.

59 T. Bencharongkul. 2022. Agoda and Bangkok Metropolitan Administration Partner Up to Drive Sustainable Growth in Bangkok with 'Urban Tree Planting' Initiative. *Agoda News*. 29 November.

60 United Nations Climate Change. 2021. The Greening of Bangkok: Can New Greening Projects Reduce Bangkok's Emissions? *Blog post*. 18 March.

61 K. Puansurin et al. 2018. The Study of Participatory Monitoring of Air Quality and Urban Heat. Case Study Udon Thani Province, Thailand. *International Journal of Agricultural Technology*. 14 (7). pp. 1693–1708.

carbon neutrality but does not specifically outline protocols for heat wave response or protection of vulnerable populations. To scale up action against heat stress, three activities could be initiated: (i) a national heat action plan framework, (ii) city- or province-specific heat or heat and air action plans, and (iii) a national fund for incentivizing and mainstreaming heat impacts in projects.

Building Blocks of a National Heat Action Plan

Figure 10 highlights the key steps the government can take to develop a national heat action plan.

Figure 10: Building Blocks of a National Heat Action Plan

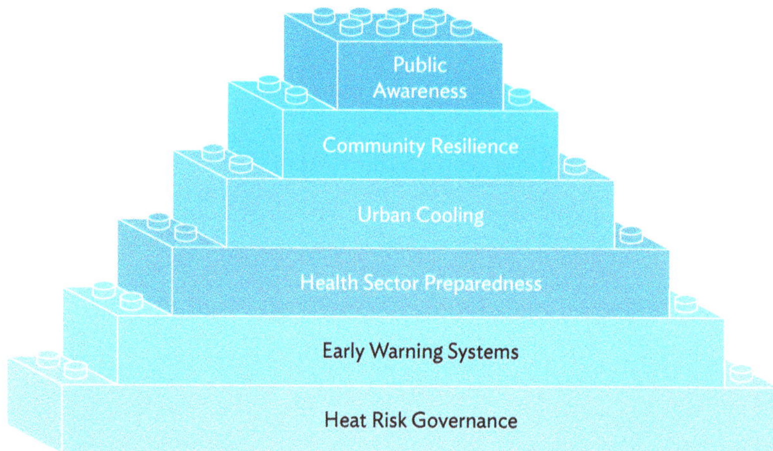

Source: Authors.

- ***Component 1: Heat Risk Governance and Systems Strengthening.*** Thailand's response to escalating heat stress begins with institutional alignment and policy coherence. This necessitates the establishment of a national heat resilience coordination unit under a suitable crosscutting ministry (e.g., health, environment, planning, etc.) to serve as the central body to guide, monitor, and synchronize heat risk interventions across ministries and provinces. This unit should work to formalize heat wave policy frameworks at all levels of governance—national, provincial, and local. Multilevel governance streamlining in design, planning, financing, and implementation is crucial for ensuring contextual relevance and local ownership. By integrating heat risks into the national adaptation plan, disaster risk reduction strategies, and urban resilience programs, the government can institutionalize heat adaptation into Thailand's broader climate and development agenda. This alignment will help unlock climate finance, streamline action plans, and support long-term resilience planning.
- ***Component 2: Early Warning Systems and Digital Infrastructure.*** To address the time-sensitive nature of extreme heat, Thailand should develop multitiered heat alert systems, mobilizing the expertise and resources of the Thai Meteorological Department. These should include location-specific thresholds and risk levels that will enable targeted alerts. Real-time dissemination, leveraging mobile apps, SMS broadcasts broadcasts, and community radio will ensure that both urban and rural populations are covered, including migrant and elder populations. Additionally, geographic information system (GIS)-based vulnerability mapping, combining satellite-derived land surface temperature with sociodemographic indicators (e.g., age, income, occupation), could support data-driven planning and resource allocation at the municipal level. Data and mapping make visible otherwise invisible marginalized populations, thereby forcing plans and budgets to take them into consideration. Stakeholder engagement and co-designing of early warning systems is very important to

understand the kind of information that channels are valued and useful to different users and vulnerable groups. This digital backbone would ensure that information not only reaches people but also informs early action and emergency response.

- ***Component 3: Health Sector Preparedness.*** Strengthening the health system's capacity to respond to heat waves is vital. Hospitals and primary care centers must be equipped with heat emergency response protocols, triage systems, and designated cooling areas. Health workers—including the *Aor Sor Mor* (village health volunteers)—must be trained in the recognition and treatment of heat-related illness, from mild dehydration to heatstroke. Furthermore, a real-time heat illness surveillance system can be rolled out, enabling rapid reporting and epidemiological tracking to inform future preparedness. This component ensures frontline health workers are ready and supported, while also improving the quality and responsiveness of care during peak heat periods.
- ***Component 4: Urban Cooling and Resilient Infrastructure***. Rapid urbanization has intensified heat exposure in Thai cities. This component could pilot nature-based and low-carbon cooling infrastructure in heat-prone urban districts. Green corridors, urban forests, and water-sensitive urban design should be introduced in pilot cities, enhancing ambient cooling while improving air quality and livability. In informal settlements and transport hubs, interventions such as heat-reflective roofing, cool pavements, solar-powered bus stops, and community misting stations can be deployed to lower surface temperatures. Public schools and hospitals should also be retrofitted with passive cooling designs, such as cross-ventilation, thermal insulation, and shaded facades—creating safer environments for vulnerable populations. Active stakeholder engagement will support the sustainability of such pilot projects.
- ***Component 5: Community Resilience and Livelihood Protection***. Communities are the first responders to heat stress, and protecting livelihoods is central to resilience. The program would establish accessible "cool zones" in public facilities like temples, libraries, and malls to offer heat relief during extreme days. With support from the Ministry of Labor, heat-safe work–rest schedules and hydration breaks should be institutionalized, especially for outdoor laborers in construction, agriculture, and street vending. Furthermore, digital safety nets and income support mechanisms—including e-cash transfers and heat index-linked insurance—can be promoted for informal workers whose productivity and health are at most risk during heat waves. These actions ensure that the poorest and most exposed communities are protected socially and economically.
- ***Component 6: Public Awareness and Behavior Change.*** A society-wide behavioral shift is critical for managing heat risks. A nationwide "Heat-Smart Thailand" campaign can drive awareness through television, social media platforms like LINE, and outdoor media. The campaign should focus on hydration, clothing, rest habits, and early signs of heat stress. Recognizing Thailand's multilingual and multicultural population, the program should produce information, education, and communication materials in multiple languages. In parallel, school and university programs can teach children and youth about heat health, self-care, and environmental responsibility—creating a heat-aware generation equipped to adapt and advocate for change.

To initiate a Heat Action Plan in Thailand, it is imperative that there is high-level engagement bringing together representatives from national, municipal, and city governments to focus on implementation, monitoring, evaluation, and data integration. Collaboration with key national ministries—such as the ministries of health, education, meteorology, and disaster management—will help define roles and trigger points for action. Additionally, financing organizations, including the Ministry of Finance, development banks like ADB, corporate social responsibility initiatives, and private sector institutions, will need to be engaged to explore potential funding sources and design financial mechanisms to support heat resilience initiatives. By uniting stakeholders and leveraging existing infrastructure, heat action planning can mitigate heat-related risks and enhance Thailand's resilience to changes in the climate. This report has curated the key requirements for a national heat action plan and presents them in the Appendix.

Pilot Heat and Air Action Plans for Cities

A key activity suggested by this report is the development of pilot city or provincial heat and air action plans. These action plans should integrate heat and air impacts—an intertwined issue in most urban areas—and develop them into an integrated system of projects and interventions. City-level heat and air action plans should be able to suggest two items:

(i) adaptation and mitigation measures, e.g., heat shelters and green spaces; and
(ii) short-term and long-term measures, e.g., retrofitting buildings in the short term and implementing district cooling networks and mass transit systems over a longer term.

Such an approach should lead to an efficient and integrated city investment plan, which can then pave the way for innovative finance structures and options.

National Heat Fund

It will also be important to consider setting up a national-level blended finance incentive fund or facility to create a real momentum for cities and provinces to develop projects with the best possible impacts on heat adaptation and mitigation, as well as provide funding for the heat and air action plans.

APPENDIX: AT A GLANCE—HEAT ACTION PLAN THAILAND

Governance and Institutional Framework	
Lead agencies	• MOPH – lead coordinator • MONRE • DDPM • TMD
Interagency coordination	• Establish a national heat resilience task force • Provincial and local government coordination
Legal and policy integration	• National Adaptation Plan • Health Security Plan • Climate Change Master Plan • Disaster risk reduction strategies
Early Warning Systems and Protocols	
Heat alert system	• Develop standardized heat index thresholds (e.g., moderate, high, extreme) • GIS-based vulnerability mapping • Community-based warning systems • SMS, television, radio, LINE, Facebook alerts
Trigger-based action protocols	• Example: “Red alert” activates school closures, outdoor work hour restrictions, etc.
Health Sector Preparedness	
Clinical protocols	• Heatstroke diagnosis and treatment guides • Hospital surge capacity plans
Health workforce training	• Train primary health workers and volunteers, e.g., *Aor Sor Mor* (village health volunteers) • Rapid response teams in high-risk areas
Public health campaigns	• Seasonal campaigns on hydration, shade, clothing, etc. • Materials in Thai, Isan, Karen, Myanmar language, and other local languages
Urban and Infrastructure Adaptation	
Urban greening	• Expand tree canopy, parks, green corridors • Water-sensitive urban design • Bangkok and secondary cities
Building and zoning codes	• Passive cooling architecture • Reflective or white roofs, heat-resistant materials • Thermal insulation
Transport infrastructure	• Solar powered bus stops • Shade and ventilation at bus and train stops • Time-of-day adjustments for roadwork

continued on next page

Appendix *continued*

Community Resilience and Livelihood Protection	
Heat vulnerability index	• Overlay social, demographic, and climatic data to identify high-risk zones
Identification of at-risk groups	• Older people • Outdoor workers (e.g., farmers, construction laborers) • Urban poor and informal settlers • Children and chronically ill
Assessment of urban heat islands	• Bangkok and secondary cities
Community cooling interventions	• "Cool zones" in temples, malls, schools, community centers • Mobile cooling vans in slums or rural areas
Heat safety in schools	• Install fans, shade cloths, trees in schoolyards • Cancel and/or restructure outdoor physical education classes during alerts
Curriculum integration	• Add heat awareness and climate health modules
Financial and social safety measures	• Utility subsidies for low-income households (electricity for fans and/or air-conditioning) • Conditional cash transfers during heat emergencies
Occupational Safety and Labor Policy	
Heat guidelines for outdoor workers	• Work–rest cycles based on wet bulb globe temperature • Access to shade, hydration, and PPE
Labor laws and enforcement	• Mandate break periods, no-work hours during peak heat • Sector focus: agriculture, construction, delivery services
Financing and Resource Mobilization	
National budget allocation	• Ring-fenced health adaptation funds under MOPH and MNRE
Donor and private sector engagement	• Green Climate Fund, ADB, UNDP, WHO • CSR partnerships for heat-resilient infrastructure
Monitoring, Evaluation, and Learning	
Data collection	• Real-time heat illness and fatality reporting system • Tracking effectiveness of interventions
Key performance indicators	• % reduction in heatstroke deaths • % of operational cooling centers • % population reached via alerts
Continuous learning	• Pilot programs in different provinces (urban, rural, coastal) • Integrate lessons into annual plan updates
Communication Strategy	
Targeted messaging	• Segment by region, occupation, age group
Communication platforms	• Use LINE, Facebook, radio, loudspeakers, and community leaders
Feedback and citizen engagement	• Two-way channels for reporting heat issues or accessing help

ADB = Asian Development Bank, CSR = corporate social responsibility, DDPM = Department of Disaster Prevention and Mitigation, GIS = geographic information system, MONRE = Ministry of Natural Resources and Environment, MOPH = Ministry of Public Health, PPE = personal protective equipment, TMD = Thai Meteorological Department, UNDP = United Nations Development Programme, WHO = World Health Organization.

Source: Authors.

www.ingramcontent.com/pod-product-compliance
Lightning Source LLC
Chambersburg PA
CBHW041800220926
51847CB00014B/266
* 9 7 8 9 2 9 2 7 7 8 4 6 0 *